文景 | Horizon

社科 新知 文艺 新潮

THE BOOK OF ANSWERS

答案之书

双语精巧版

CAROL BOLT

[加拿大]卡罗尔·博尔特 著

何静 译

上海人民出版社

6.44   It is not how things are in the world that is mystical, but that it exists.

6.44   神秘的不是世界如何，神秘的是世界存在。

—— 路德维希·维特根斯坦，《逻辑哲学论》

湖 岸

Hu'an publications®

如何使用本书

用 10~15 秒集中精力思考你的问题
问题必须是一句完整的话
例如:"这项新的工作好吗?"
"我会幸福吗?"等等

拿起本书,翻动书页
默念或说出你想要问的问题
(每次只问一个)

停在感觉合适的时间
打开翻到的那一页
你就会找到问题的答案

YOUR ACTIONS WILL

IMPROVE THINGS

你的行动会使事情变得更好

DON'T BET ON IT

那可不一定

ADOPT
AN ADVENTUROUS ATTITUDE

采取积极进取的态度

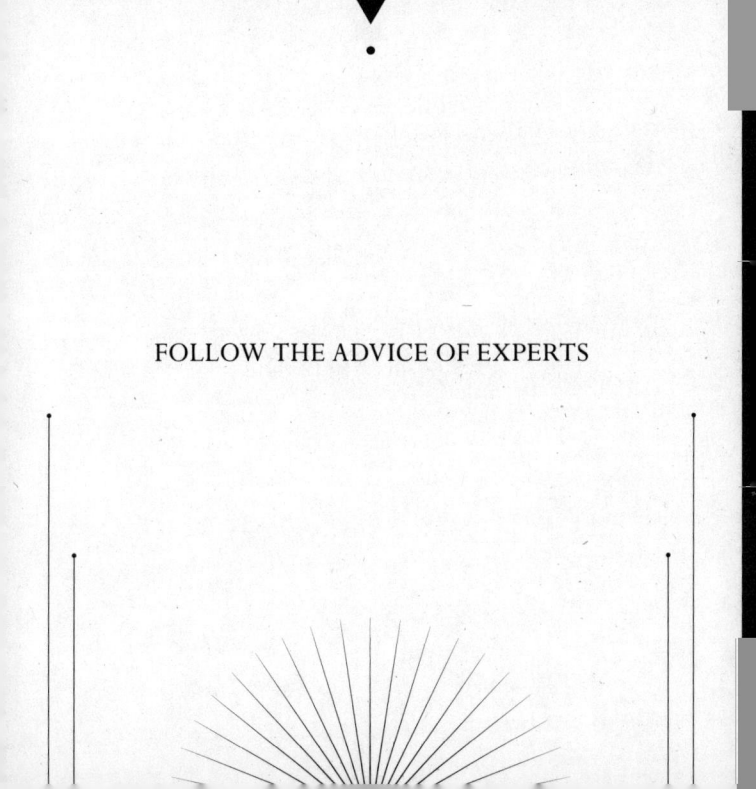

FOLLOW THE ADVICE OF EXPERTS

遵循专家的建议

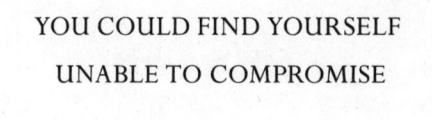

YOU COULD FIND YOURSELF

UNABLE TO COMPROMISE

你会发现自己无法妥协

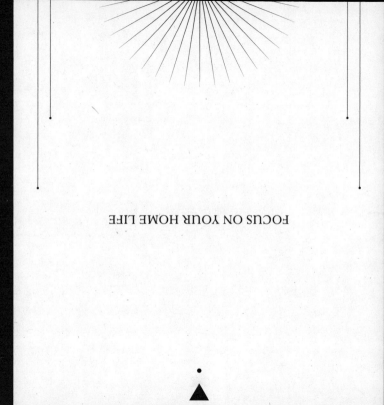

FOCUS ON YOUR HOME LIFE

专注于你的家庭生活

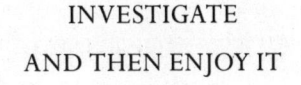

INVESTIGATE

AND THEN ENJOY IT

探索并享受它

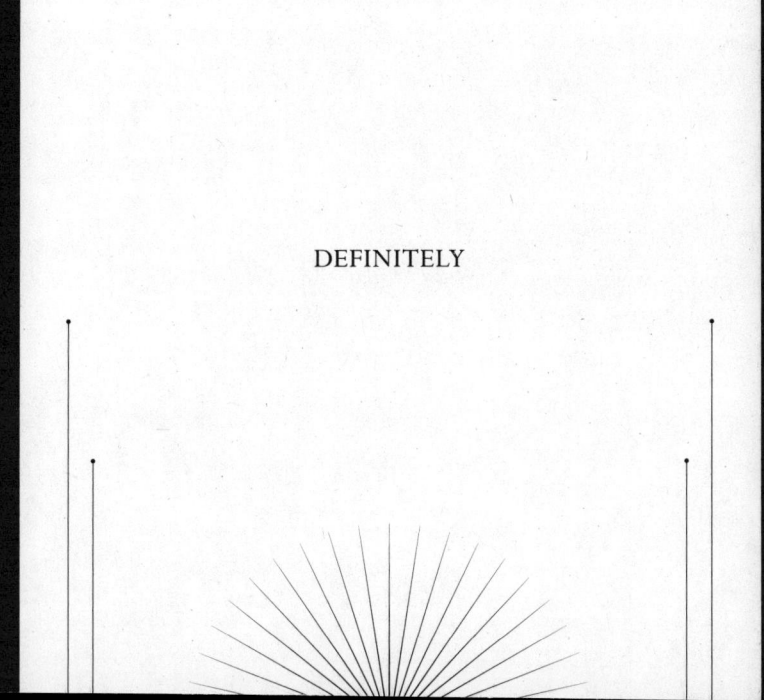

DEFINITELY

肯定地

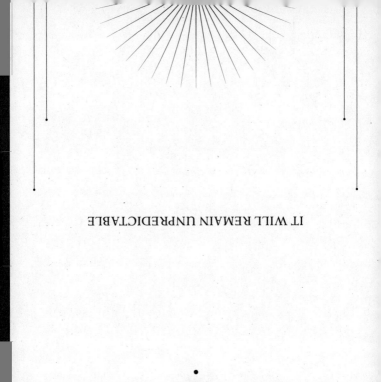

IT WILL REMAIN UNPREDICTABLE

仍然难以预料

ABSOLUTELY NOT

绝对不是

EXPLORE IT

WITH PLAYFUL CURIOSITY

好奇地去探索

BE DELIGHTFULLY SURE OF IT

可以非常高兴地确定

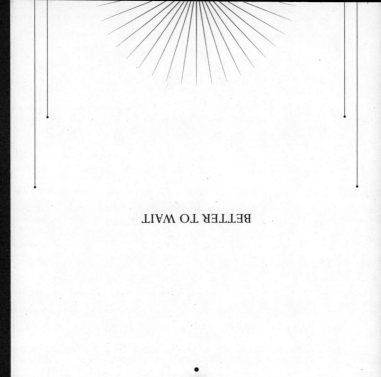

BETTER TO WAIT

最好等等

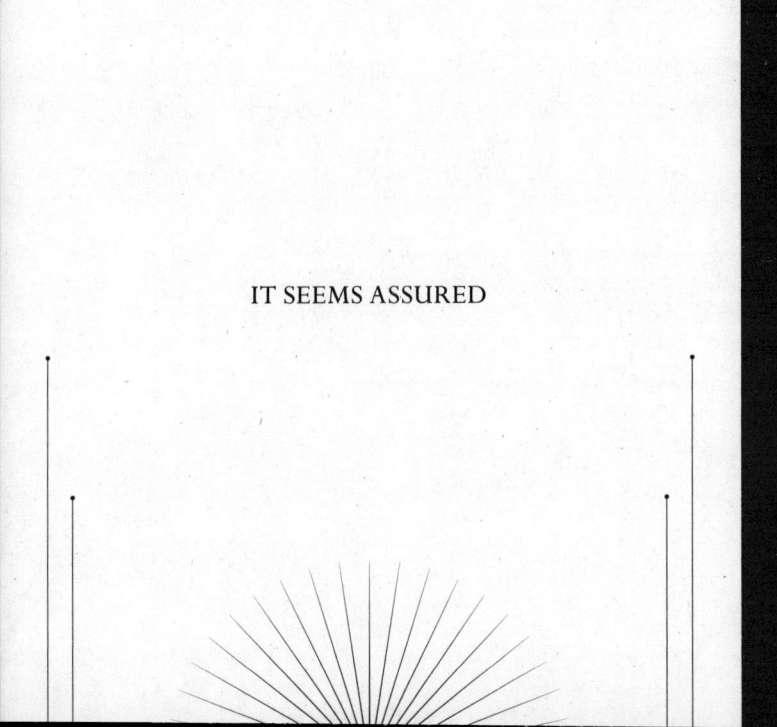

IT SEEMS ASSURED

似乎没有问题

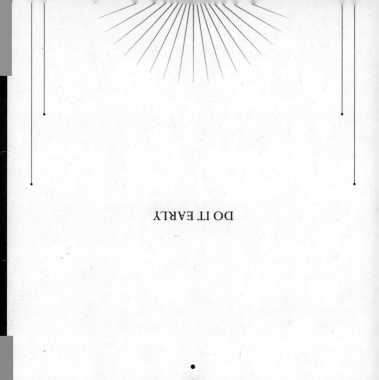

DO IT EARLY

尽早行动

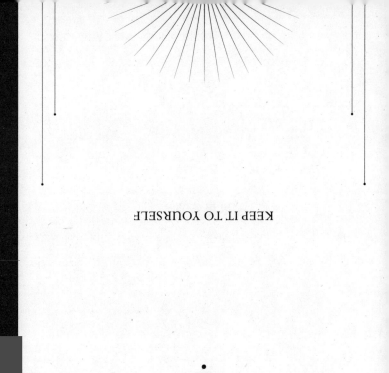

KEEP IT TO YOURSELF

不要告诉别人

STARTLING EVENTS

MAY OCCUR AS A RESULT

结果可能会令人吃惊

THE ANSWER MAY COME TO YOU

IN ANOTHER LANGUAGE

答案可能会以另一种形式出现

YOU WILL NEED
TO ACCOMMODATE

你需要适应

DOUBT IT

怀疑

IT WILL BRING GOOD LUCK

它会带来好运

BE PATIENT

要有耐心

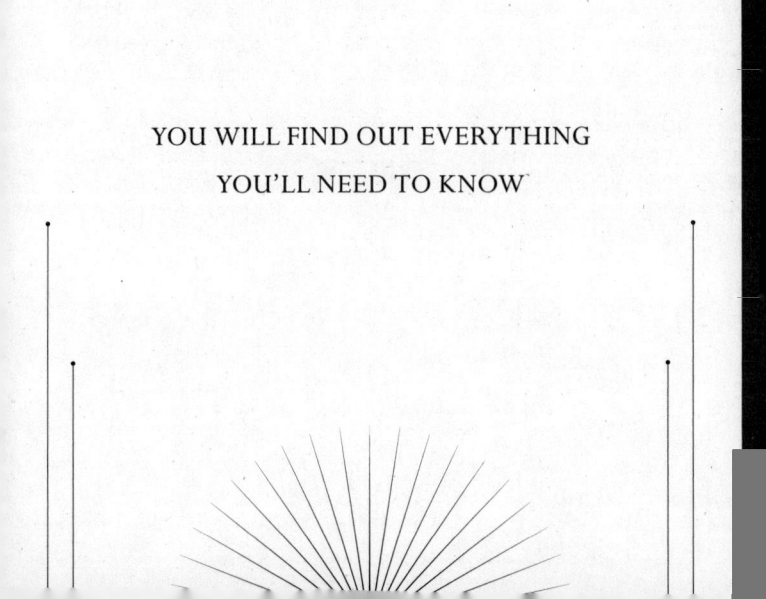

YOU WILL FIND OUT EVERYTHING
YOU'LL NEED TO KNOW

需要知道的, 你都会知道

THERE IS A SUBSTANTIAL
LINK TO ANOTHER SITUATION

与另一种情况大有关系

WATCH AND SEE WHAT HAPPENS

观察看看会发生什么

IT WILL AFFECT

HOW OTHERS SEE YOU

这会影响其他人对你的看法

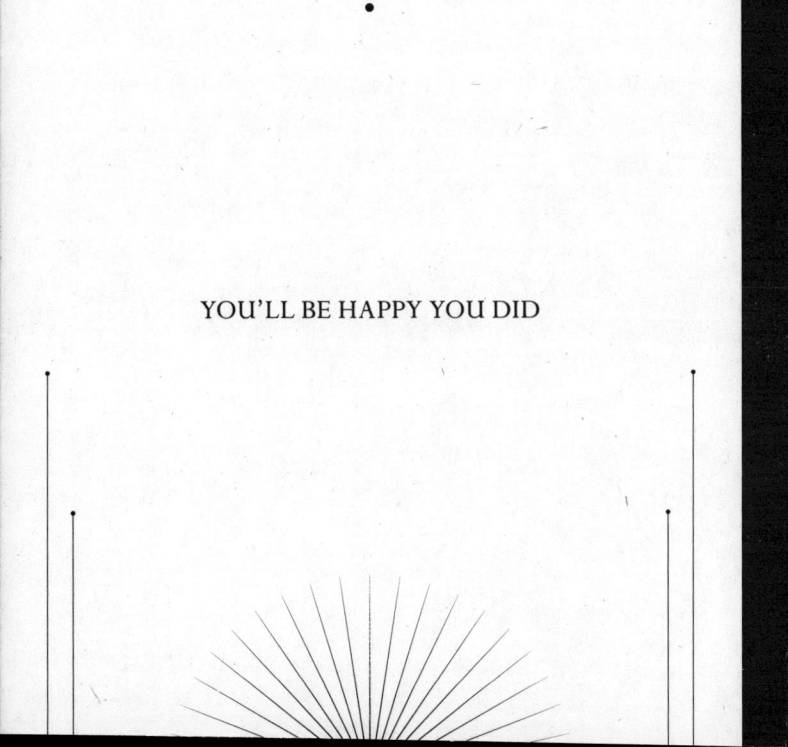

YOU'LL BE HAPPY YOU DID

你会高兴你做了

GET IT IN WRITING

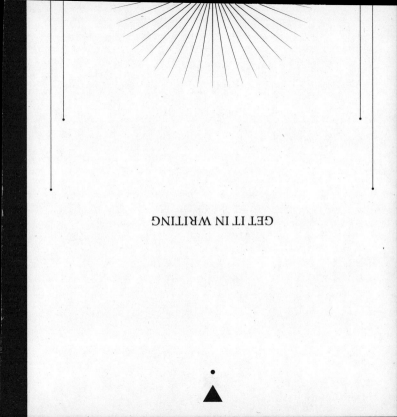

写下来

UNFAVORABLE AT THIS TIME

此时不宜

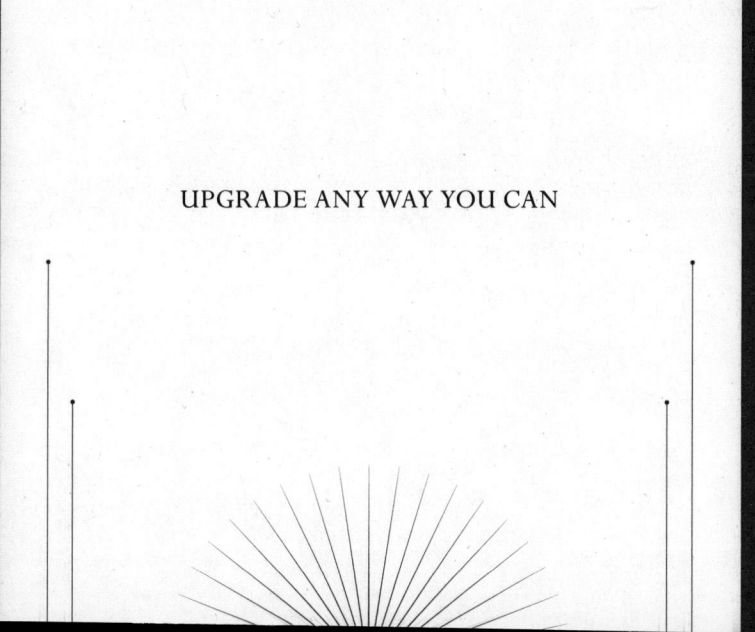

UPGRADE ANY WAY YOU CAN

尽力提高

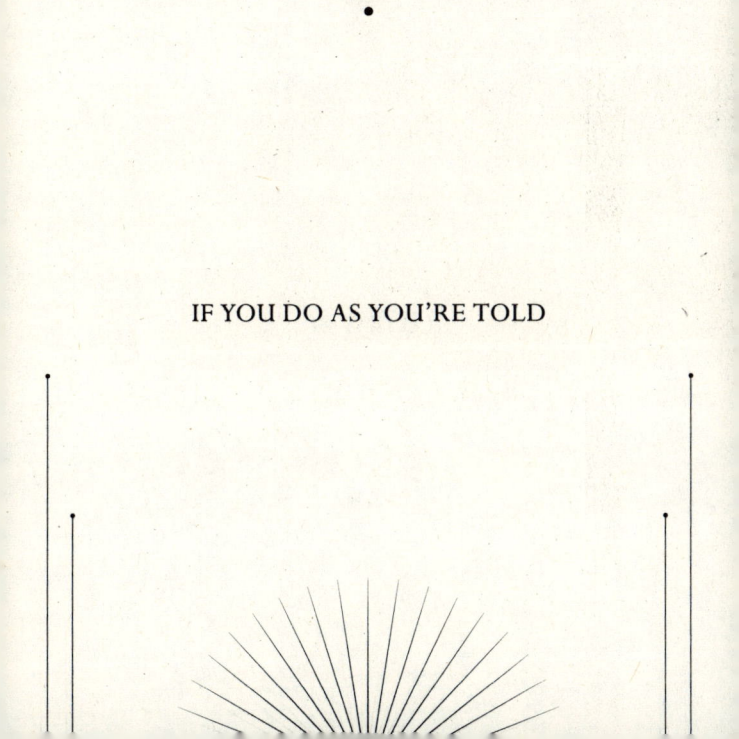

IF YOU DO AS YOU'RE TOLD

只要你按照被告知的方法去做

IF IT'S DONE WELL,

IF NOT, DON'T DO IT AT ALL

能做好就做，否则干脆别做

DON'T ASK FOR ANYMORE
AT THIS TIME

此时不要要求更多

AVOID THE FIRST SOLUTION

不要想到什么就做什么

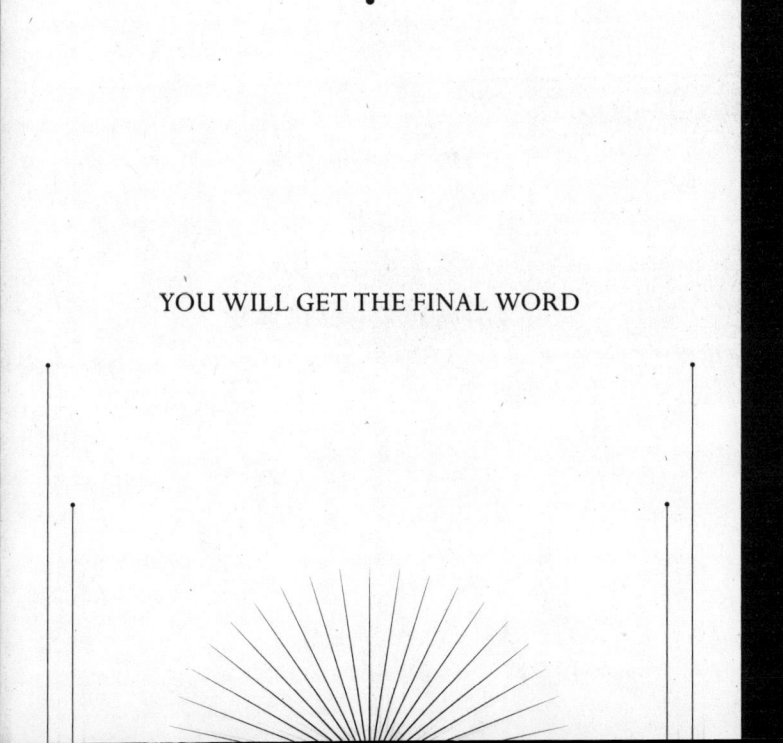

YOU WILL GET THE FINAL WORD

将由你来决定

PROCEED AT A MORE

RELAXED PACE

慢慢来

THE BEST SOLUTION

MAY NOT BE THE OBVIOUS ONE

轻易想到的办法未必是最好的

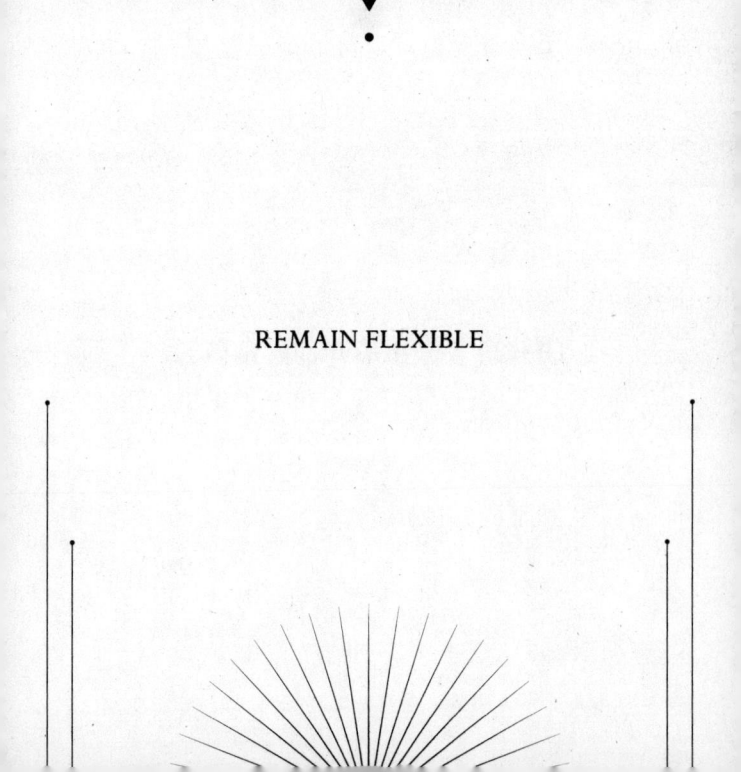

REMAIN FLEXIBLE

保持变通

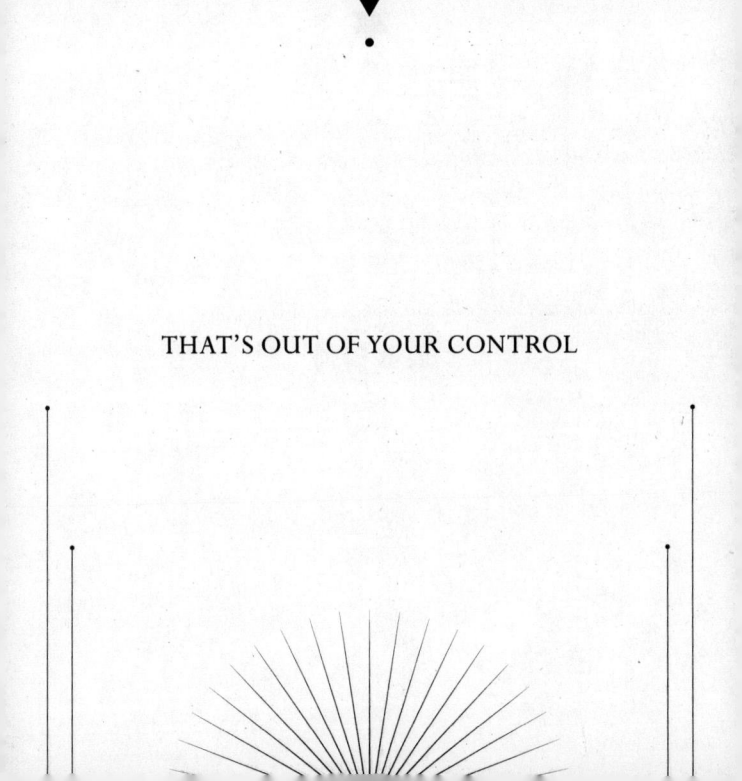

THAT'S OUT OF YOUR CONTROL

超出你的掌控范围

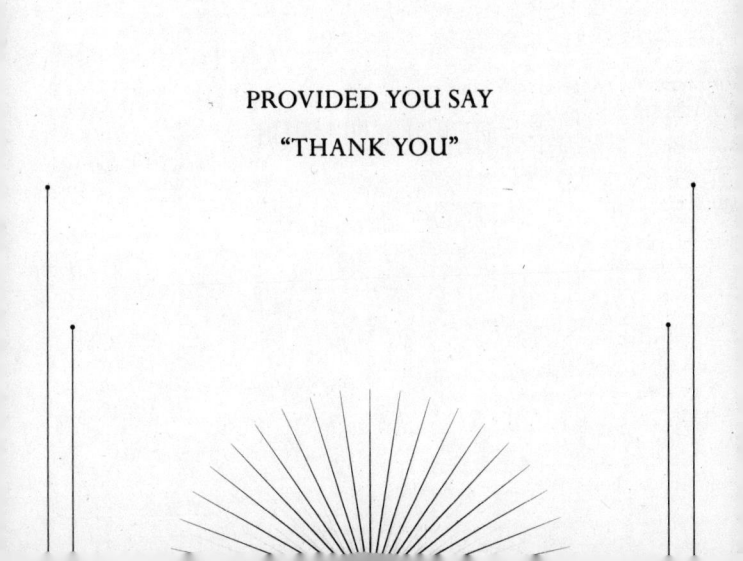

PROVIDED YOU SAY

"THANK YOU"

只要你说过"谢谢"

ENJOY THE EXPERIENCE

享受这种经历

APPROACH CAUTIOUSLY

谨慎行事

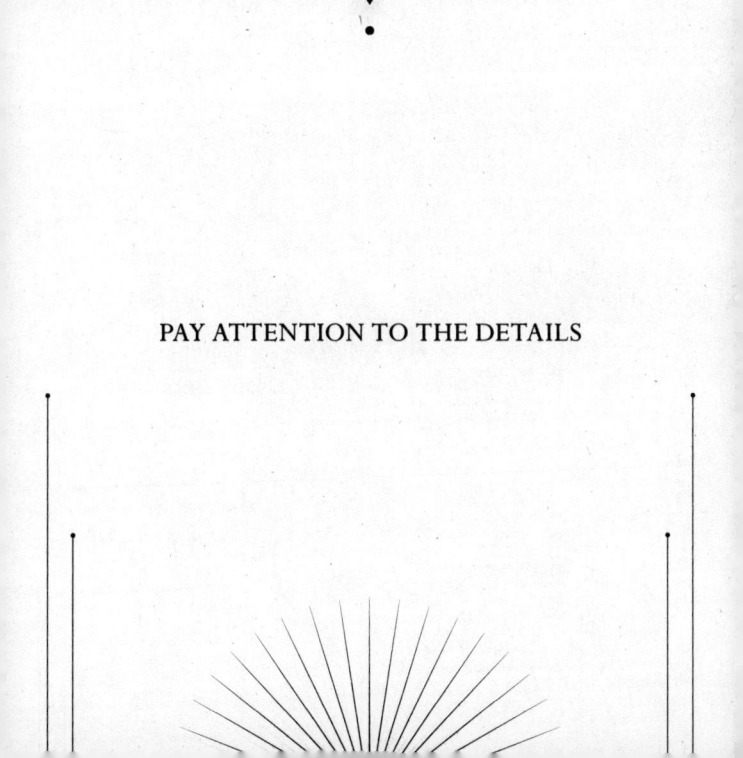

PAY ATTENTION TO THE DETAILS

注意细节

WATCH YOUR STEP AS YOU GO

留神走的每一步

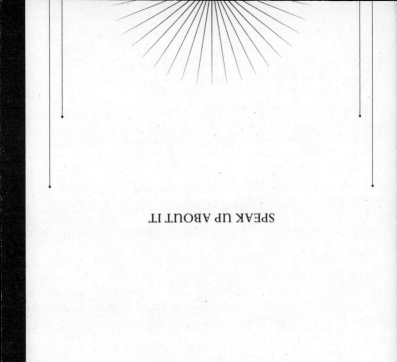

SPEAK UP ABOUT IT

大声说出来

DO NOT HESTITATE

不要犹豫

THIS IS A GOOD TIME TO

MAKE A NEW PLAN

此时是制订新计划的好时机

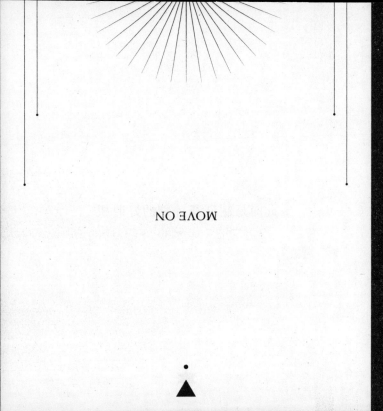

MOVE ON

放下，然后继续

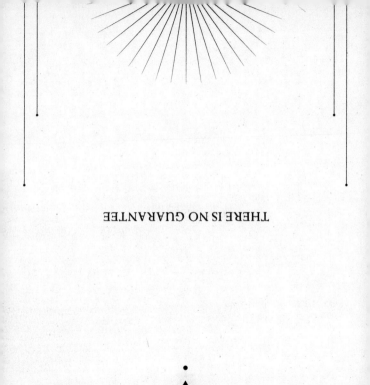

THERE IS NO GUARANTEE

无法保证

THE CIRCUMSTANCES
WILL CHANGE VERY QUICKLY

情况很快就会有所改变

DO NOT GET CAUGHT UP
IN YOUR EMOTIONS

不要被情绪左右

SHIFT YOUR FOCUS

转移你的焦点

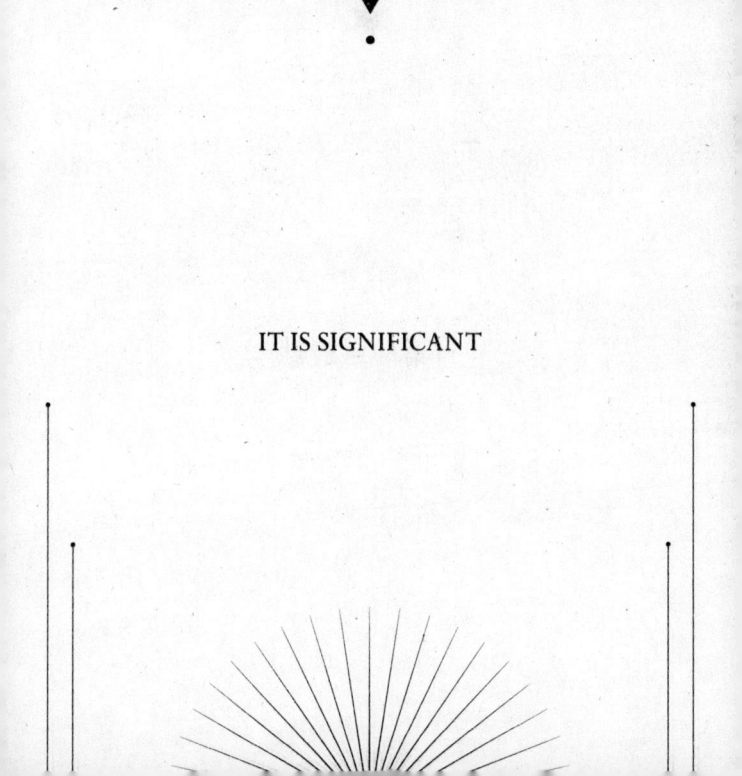

IT IS SIGNIFICANT

极为重要

REPRIORITIZE
WHAT IS IMPORTANT

弄清事情的轻重缓急

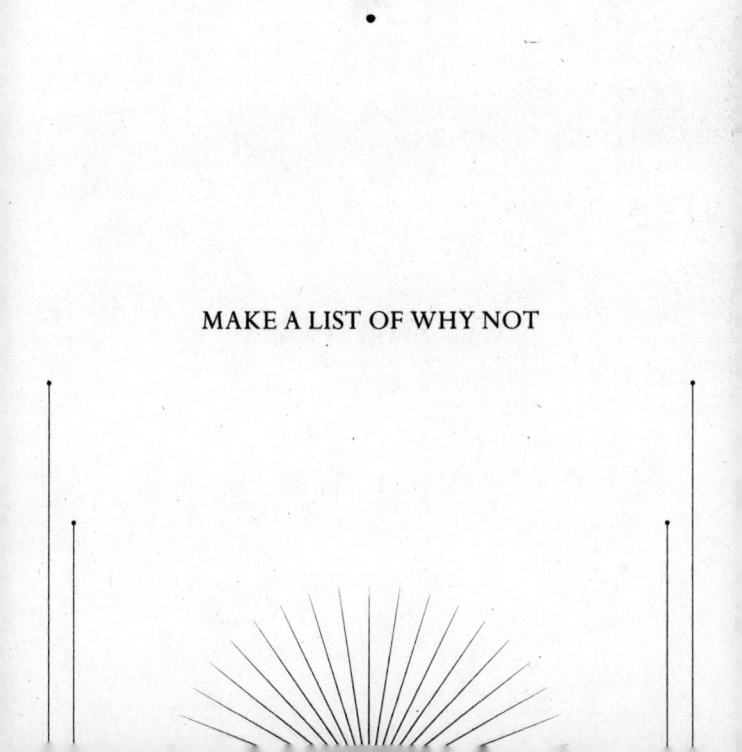

MAKE A LIST OF WHY NOT

列出不做的理由

DO NOT WAIT

不要等待

IT IS SOMETHING

YOU WON'T FORGET

将使你难忘

EXPECT TO SETTLE

期待解决

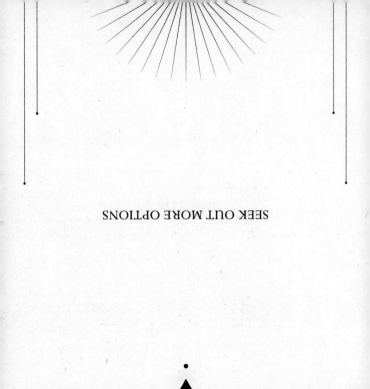

SEEK OUT MORE OPTIONS

寻求更多的选择

FOLLOW THROUGH
ON YOUR OBLIGATIONS

履行你的义务

DEAL WITH IT LATER

先不做决定

FOLLOW SOMEONE ELSE'S LEAD

跟着其他人走

MAKE A LIST OF WHY

列出做的理由

TAKE A CHANCE

冒险一试

ACCEPT A CHANGE
TO YOUR ROUTINE

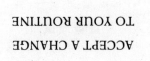

接受新的变化

YOU WILL NEED TO TAKE

THE INITIATIVE

你需要争取主动

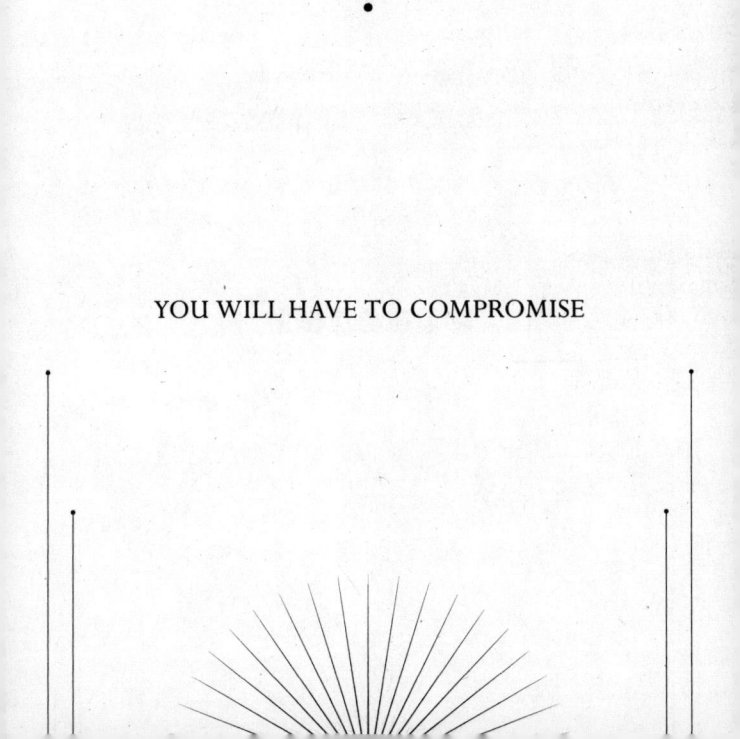

YOU WILL HAVE TO COMPROMISE

你必须妥协

YOU WILL NEED
MORE INFORMATION

你需要更多的信息

TRUST
YOUR ORIGINAL THOUGHT

相信你最初的想法

IT WILL CREATE A STIR

它会引起轰动

REMOVE YOUR OWN OBSTACLES

排除你自己的障碍

IT WOULD BE BETTER

TO FOCUS ON YOUR WORK

最好专注于你的工作

IT WILL BE A PLEASURE

将会很愉快

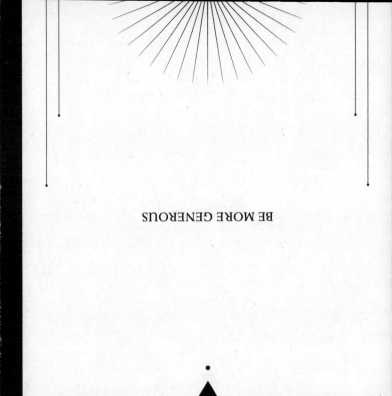

BE MORE GENEROUS

慷慨一点

BET ON IT

十拿九稳

FINISH SOMETHING ELSE FIRST

先完成其他的事情

YOU MAY HAVE OPPOSITION

你可能遭到反对

YOU ARE TOO CLOSE TO SEE

你离得太近，难以看清

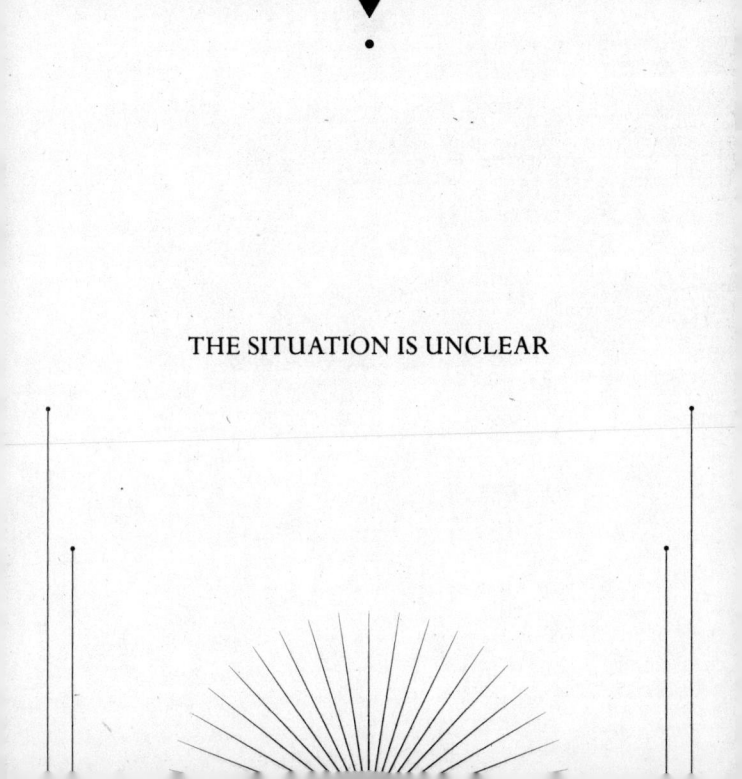

THE SITUATION IS UNCLEAR

情况不明

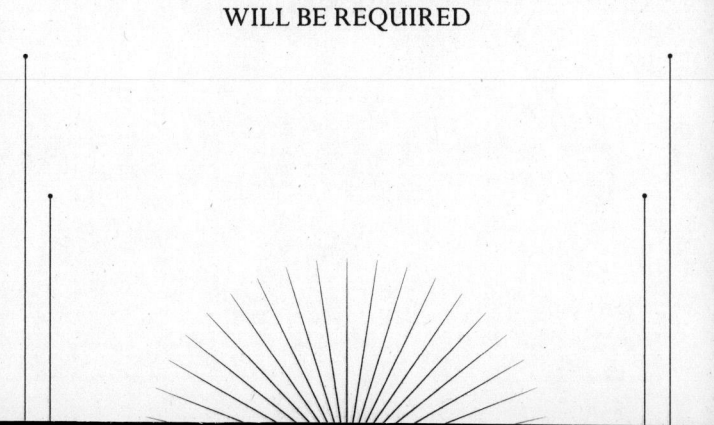

A SUBSTANTIAL EFFORT

WILL BE REQUIRED

需要付出巨大努力

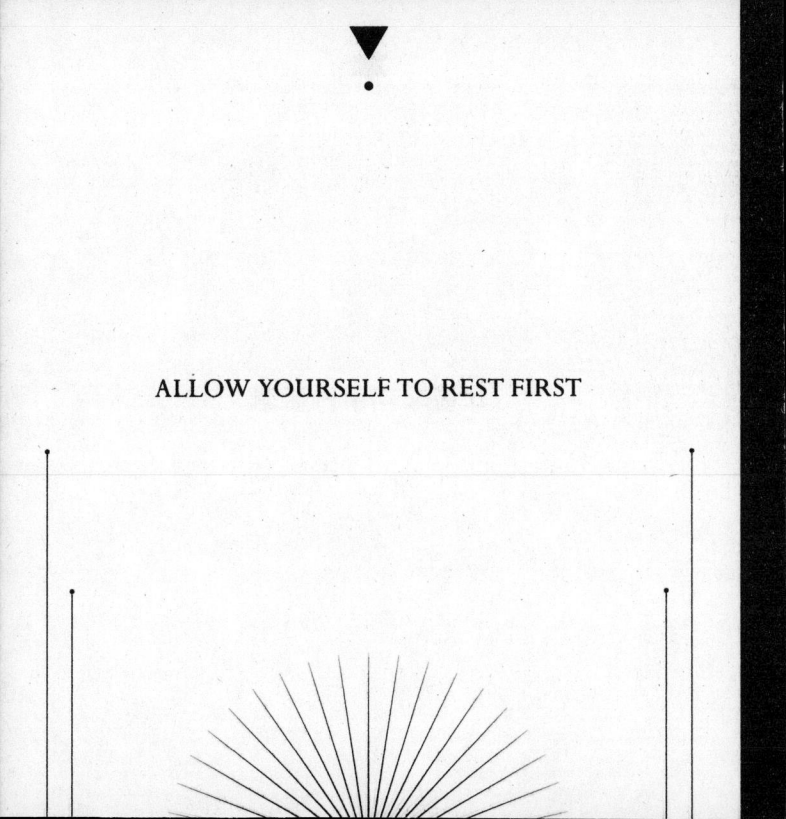

ALLOW YOURSELF TO REST FIRST

让自己先休息一下

THE CHANCE
WILL NOT COME AGAIN SOON

这个机会很难得

重新考虑你的方法

IT WOULD BE INADVISABLE

不妥

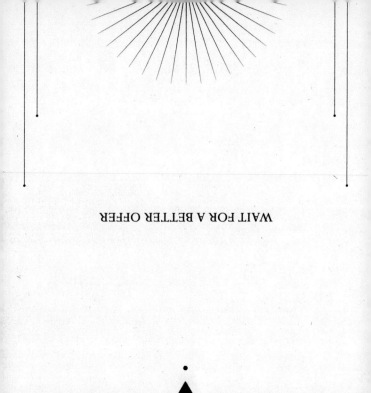

WAIT FOR A BETTER OFFER

等待更好的机会

SETTLE IT SOON

尽快解决

YES,
BUT DON'T FORCE IT

是的，但不要强迫

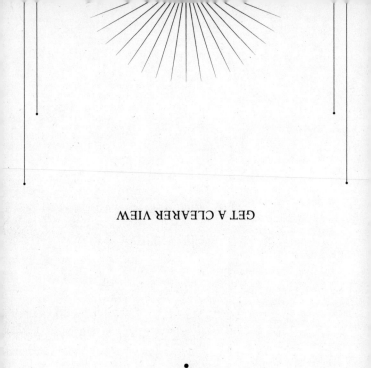

GET A CLEARER VIEW

看得更清楚一点

TAKE A CHANCE

冒险一试

NOW YOU CAN

现在可以了

DON'T OVERDO IT

不可过分

IT WILL SUSTAIN YOU

它会让你受益

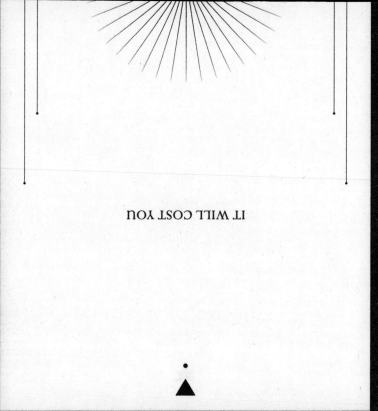

IT WILL COST YOU

它会使你付出

IT IS SURE TO MAKE THINGS
INTERESTING

这会使事情变得更有趣

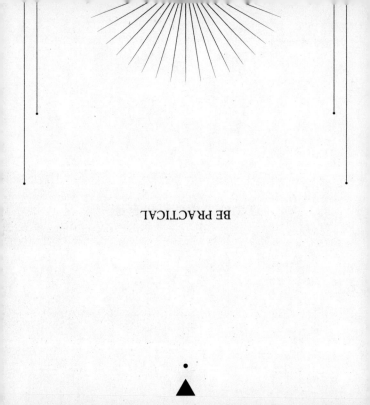

BE PRACTICAL

实际点

SAVE YOUR ENERGY

保存你的精力

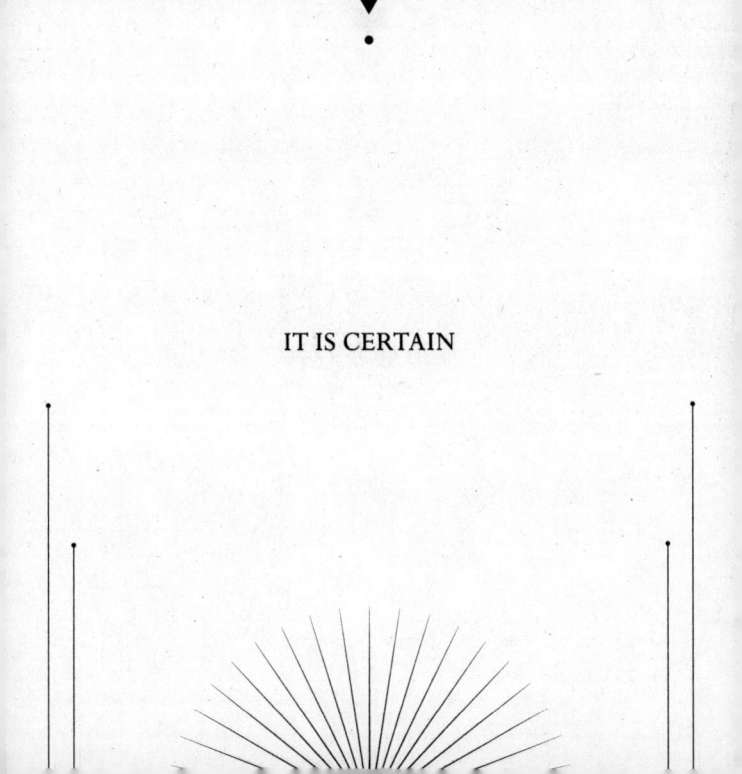

IT IS CERTAIN

确定无疑

IT IS UNCERTAIN

不确定

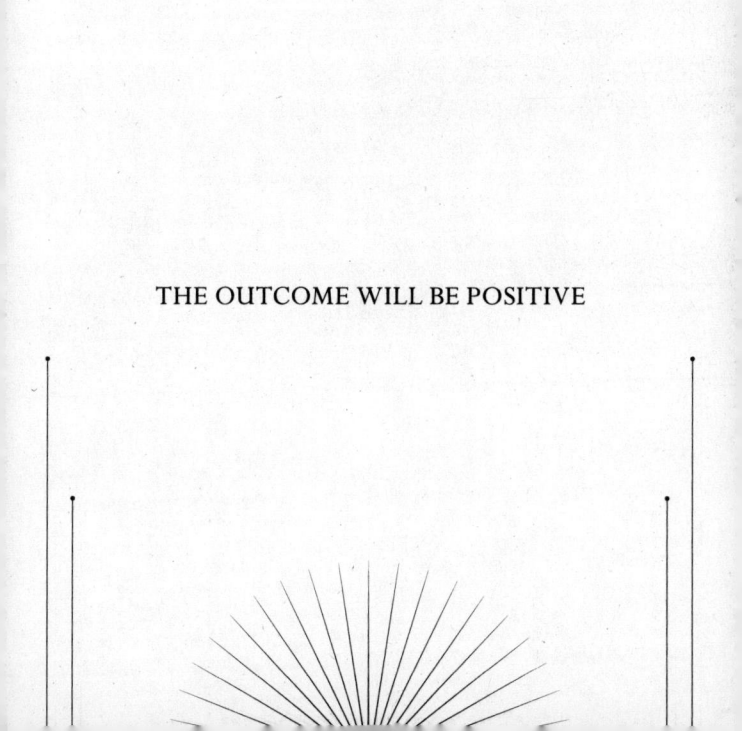

THE OUTCOME WILL BE POSITIVE

结果将是好的

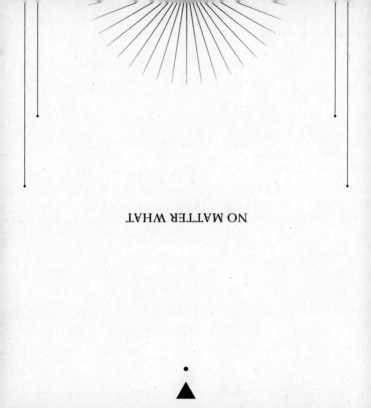

NO MATTER WHAT

哪怕天上下刀子

YOU MAY HAVE TO DROP
OTHER THINGS

可能不得不放弃其他事情

DO NOT BE CONCERNED

不要担心

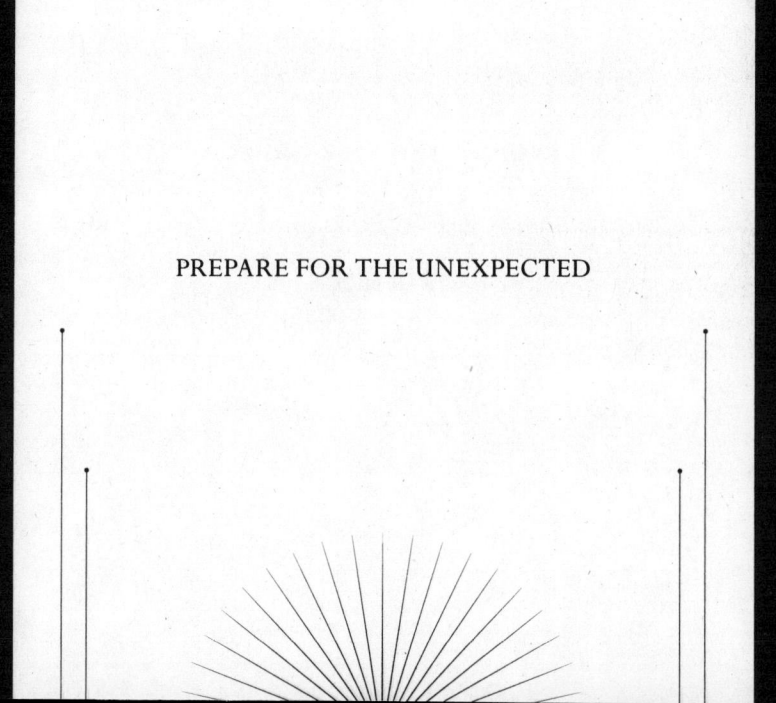

PREPARE FOR THE UNEXPECTED

未雨绸缪

IT IS NOT SIGNIFICANT

这不重要

TELL SOMEONE
WHAT IT MEANS TO YOU

告诉别人，这对你意味着什么

WHATEVER YOU DO

THE RESULTS WILL BE LASTING

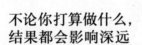

不论你打算做什么，
结果都会影响深远

KEEP AN OPEN MIND

保持开放的态度

IT IS A GOOD TIME

TO MAKE PLANS

这是制订计划的好时机

IT MAY BE DIFFICULT
BUT YOU WILL FIND VALUE IN IT

有难度，但值得做

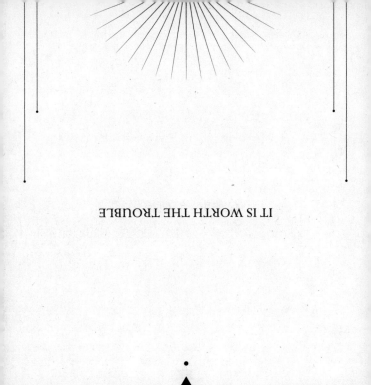

IT IS WORTH THE TROUBLE

值得去做

THERE WILL BE OBSTACLES

TO OVERCOME

会有需要克服的障碍

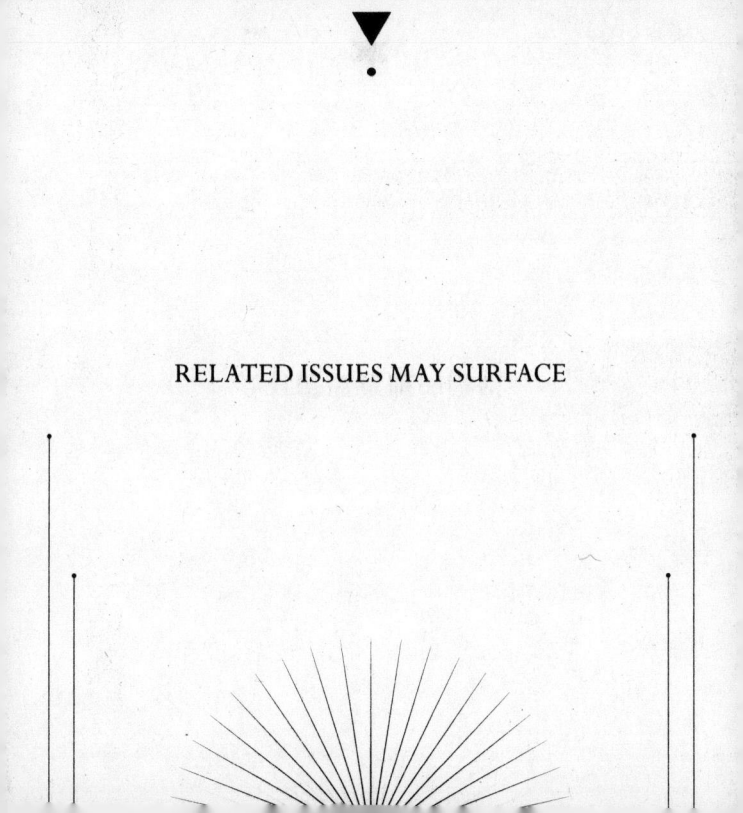

RELATED ISSUES MAY SURFACE

相关问题可能会浮出水面

YOU ARE SURE

TO HAVE SUPPORT

你一定能获得支持

ASSISTANCE WOULD MAKE

YOUR PROGRESS

A SUCCESS

可获得帮助，并能成功

COLLABORATION
WILL BE KEY

合作是关键

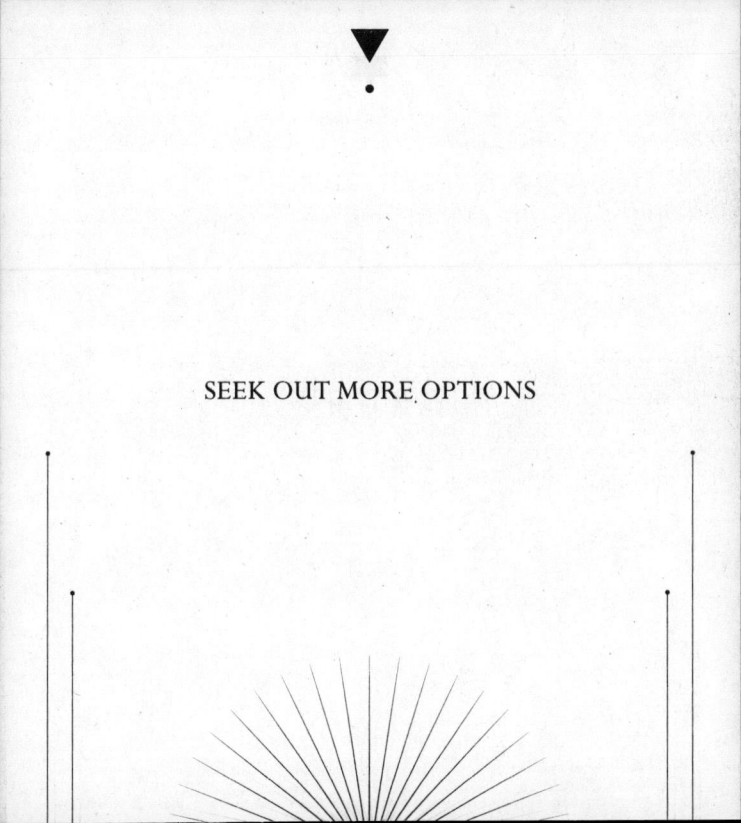

SEEK OUT MORE OPTIONS

寻求更多的选择

TAKE CHARGE

担起责任

IT CANNOT FAIL

不能失败

YOU MUST ACT NOW

你现在必须行动

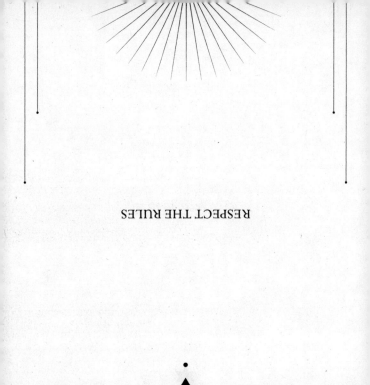

RESPECT THE RULES

遵守规则

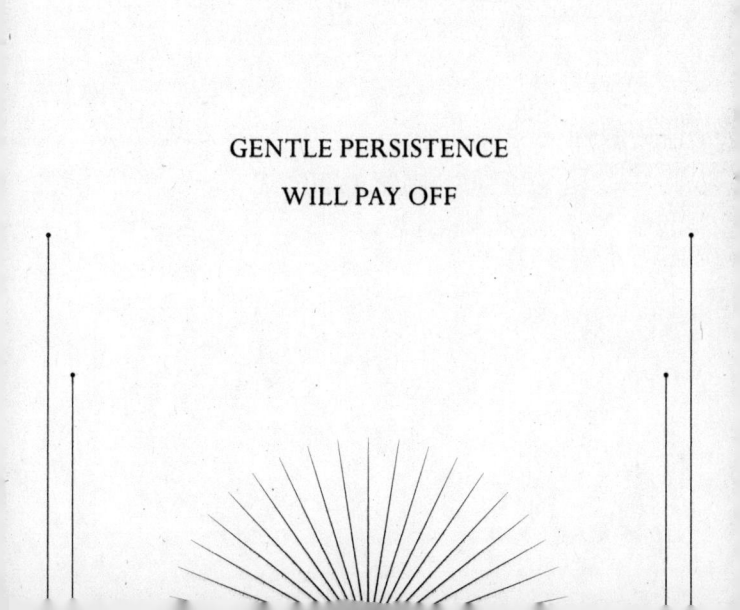

GENTLE PERSISTENCE

WILL PAY OFF

坚持一定会有回报

YOU WILL NOT BE DISAPPOINTED

不会令你失望

IT MAY ALREADY BE

A DONE DEAL

可能已成定局

FOLLOW THROUGH
WITH YOUR GOOD INTENTIONS

坚持你的善意

TAKE MORE TIME TO DECIDE

不急于决定

DO NOT BE PRESSURED

INTO ACTING TOO QUICKLY

不要迫于压力而仓促行事

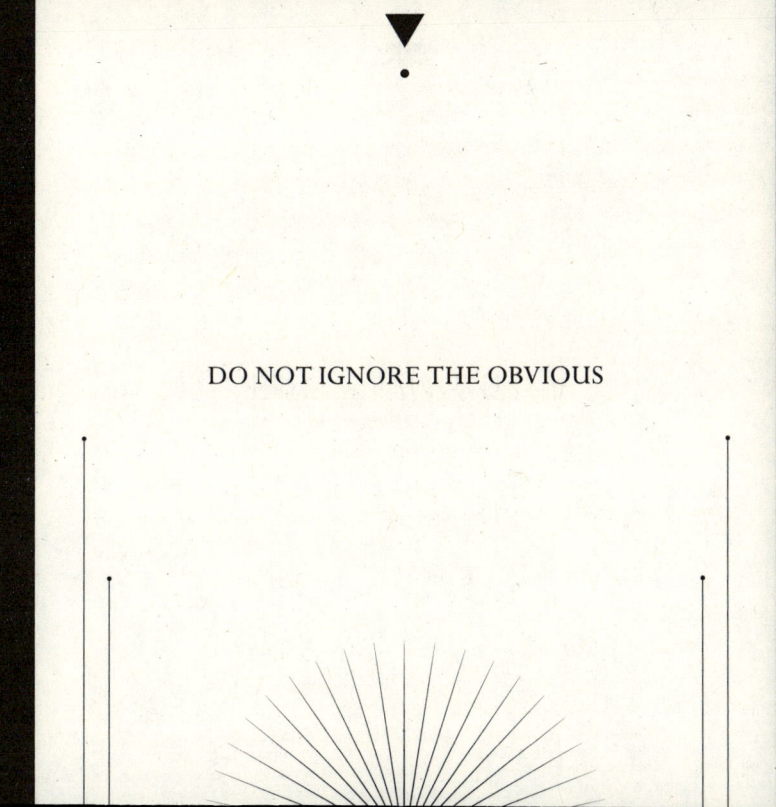

DO NOT IGNORE THE OBVIOUS

不要忽略明显的事实

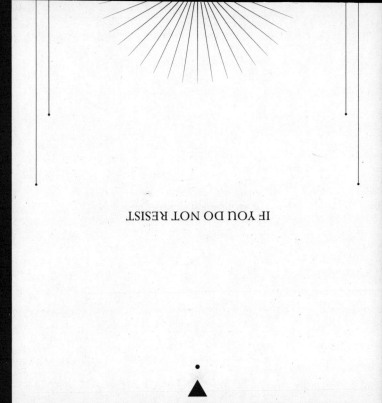

IF YOU DO NOT RESIST

如果你不抗拒的话

IT IS NOT WORTH A STRUGGLE

不值一搏

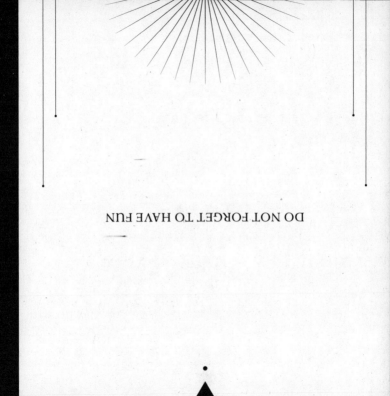

DO NOT FORGET TO HAVE FUN

别忘了让自己开心点

DO NOT DOUBT IT

不要怀疑

A STRONG COMMITMENT

WILL ACHIEVE GOOD RESULTS

全心投入会取得好的结果

TRY A MORE UNLIKELY SOLUTION

尝试一种不太可能想到的解决方案

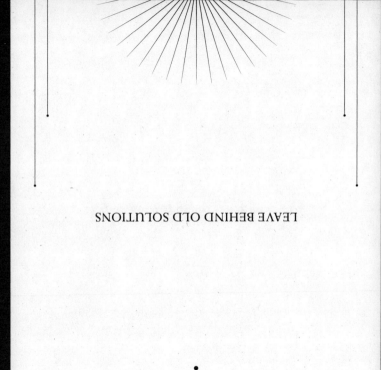

LEAVE BEHIND OLD SOLUTIONS

放弃旧的解决方案

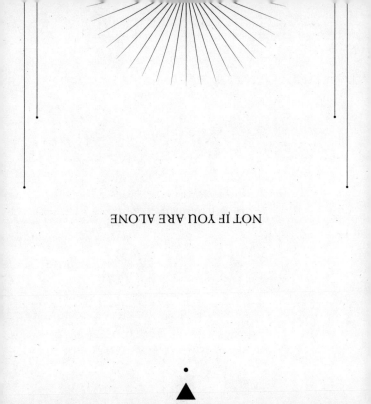

NOT IF YOU ARE ALONE

如果你一个人，就不要

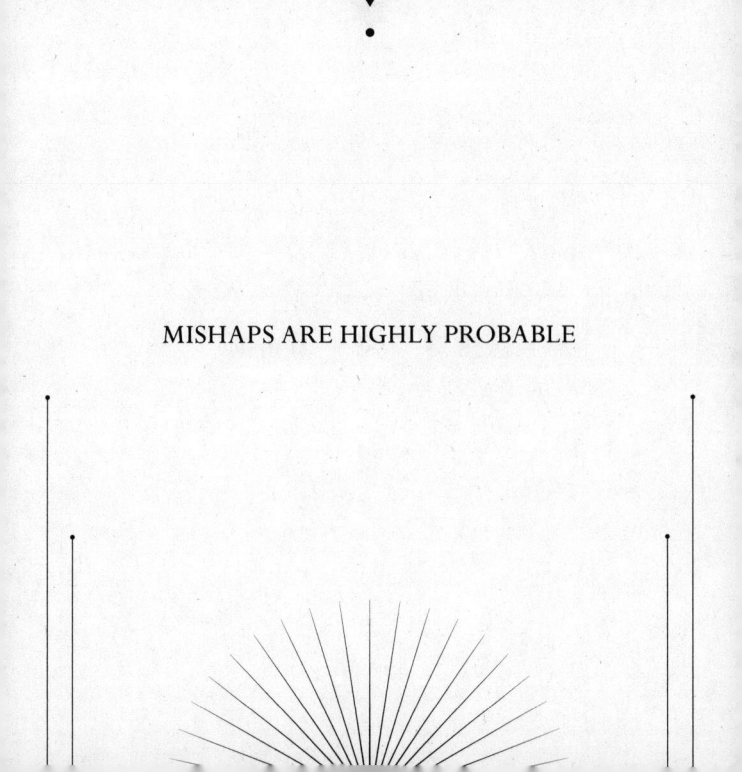

MISHAPS ARE HIGHLY PROBABLE

极有可能出现意外状况

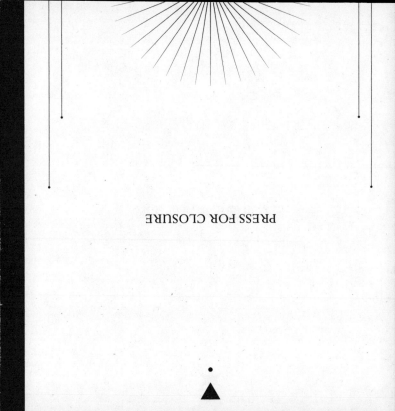

PRESS FOR CLOSURE

尽快了结

REALIZE THAT

TOO MANY CHOICES IS AS

DIFFICULT AS TOO FEW

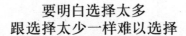

要明白选择太多
跟选择太少一样难以选择

YES

是的

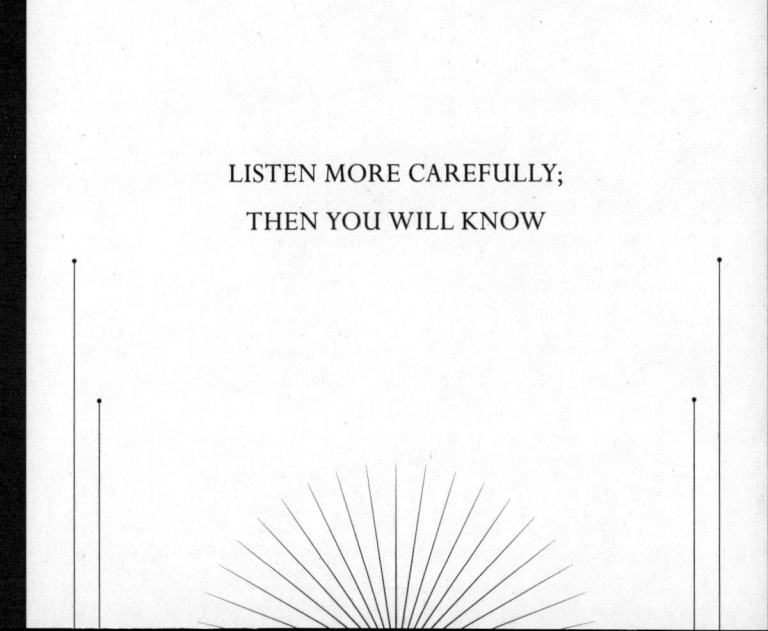

LISTEN MORE CAREFULLY;

THEN YOU WILL KNOW

仔细听，便会明白

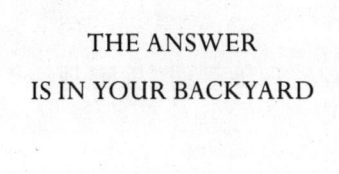

THE ANSWER

IS IN YOUR BACKYARD

答案就在你内心深处

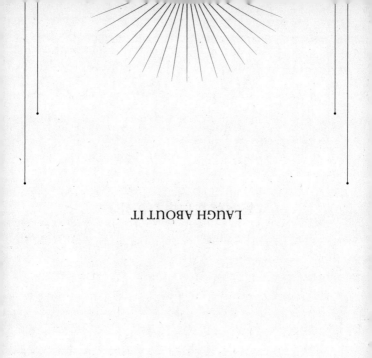

LAUGH ABOUT IT

一笑置之

OTHERS WILL DEPEND

ON YOUR CHOICES

其余的，取决于你的选择

LET IT GO

随它吧

THAT WOULD BE A WASTE

OF MONEY

浪费钱

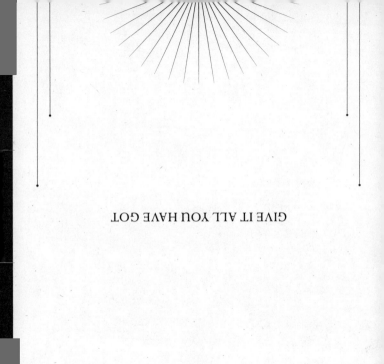

GIVE IT ALL YOU HAVE GOT

竭尽全力

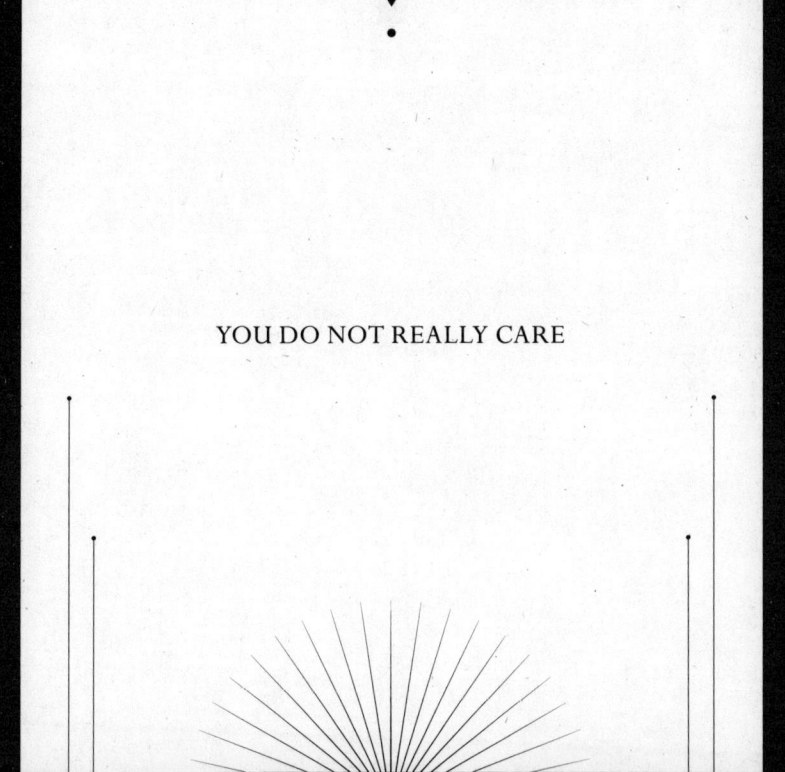

YOU DO NOT REALLY CARE

你并非真的在意

YOU WILL NEED TO

CONSIDER OTHER WAYS

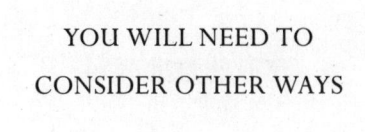

你需要考虑其他办法

A YEAR FROM NOW
IT WILL NOT MATTER

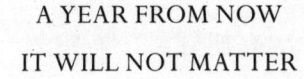

一年之后这将不再重要

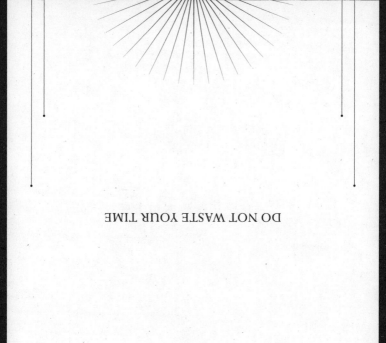

DO NOT WASTE YOUR TIME

不要浪费你的时间

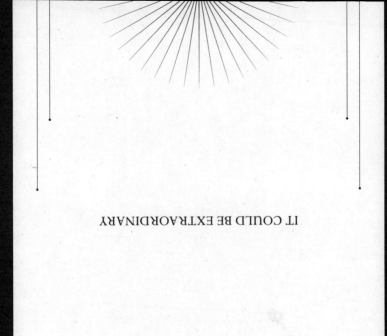

IT COULD BE EXTRAORDINARY

可能会异乎寻常地好

COUNT TO 10;
ASK AGAIN

从 1 数到 10，再问一次

ACT AS THOUGH
IT IS ALREADY REAL

就当它是既成事实

SETTING PRIORITIES

WILL BE A NECESSARY PART

OF THE PROCESS

弄清轻重缓急是此事的重中之重

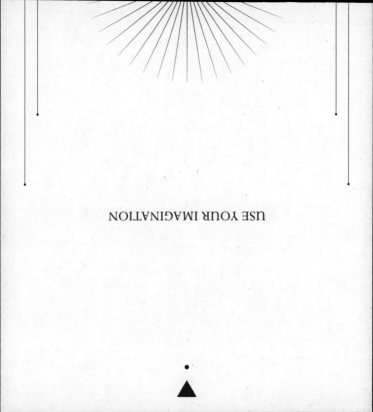

USE YOUR IMAGINATION

发挥你的想象力

IT IS GONNA BE GREAT

一定会非常成功

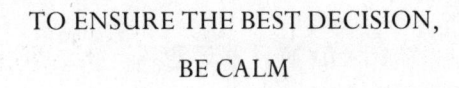

TO ENSURE THE BEST DECISION,

BE CALM

冷静才能做出最好的选择

WAIT

等待

YOU WILL HAVE TO MAKE IT UP

AS YOU GO

你到时将不得不随机应变

YOU WILL REGRET IT

你会后悔的

UNQUESTIONABLY

毫无疑问

OF COURSE

当然

YOU KNOW BETTER NOW
THAN EVER BEFORE

你现在比以前任何时候都更清楚

TRUST YOUR INTUITION

相信你的直觉

CONSIDER IT AN OPPORTUNITY

把它看作一个机会

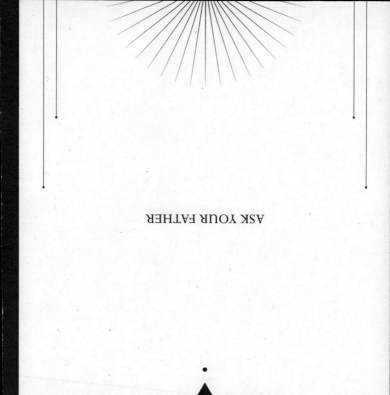

ASK YOUR FATHER

问问父亲

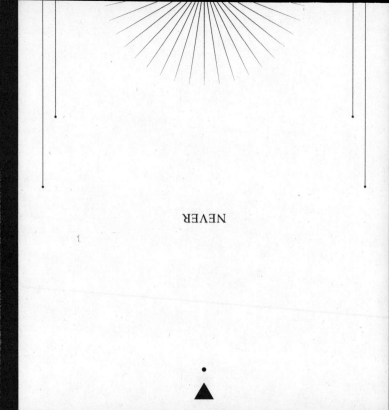

NEVER

绝不

ASK YOUR MOTHER

问问母亲

PERHAPS, WHEN YOU ARE OLDER

或许吧，等你再长大一点

ONLY DO IT ONCE

仅此一次

MAYBE

也许吧

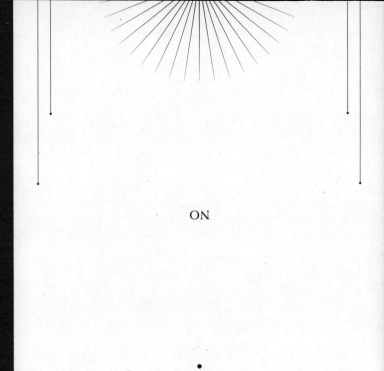

ON

不

YES

是的

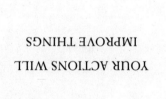

YOUR ACTIONS WILL
IMPROVE THINGS

你的行动会使事情变得更好

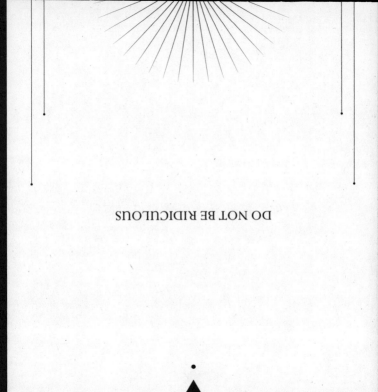

DO NOT BE RIDICULOUS

别傻了

DON'T BET ON IT

那可不一定

ADOPT
AN ADVENTUROUS ATTITUDE

采取积极进取的态度

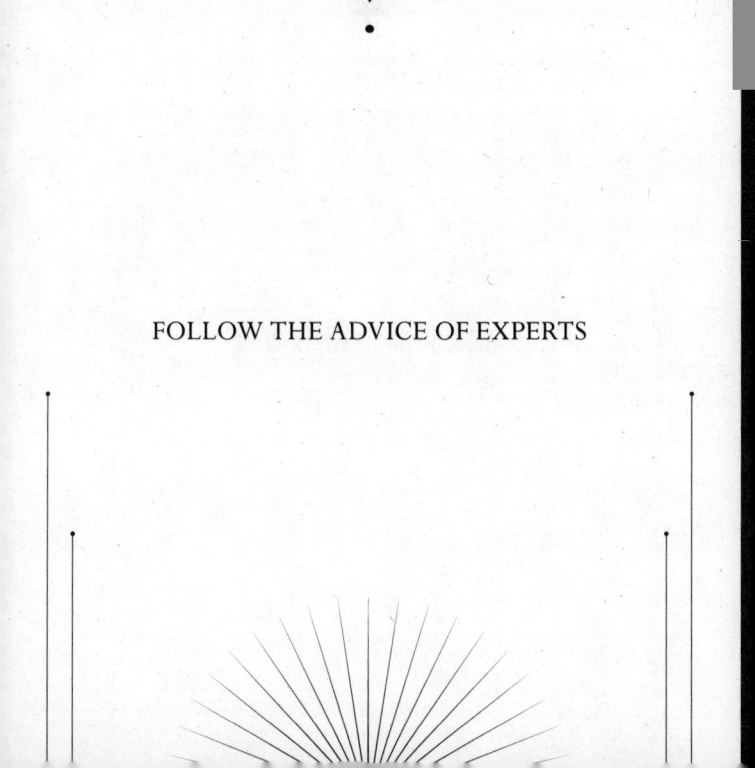

FOLLOW THE ADVICE OF EXPERTS

遵循专家的建议

YOU COULD FIND YOURSELF

UNABLE TO COMPROMISE

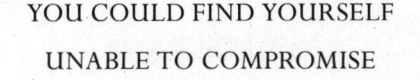

你会发现自己无法妥协

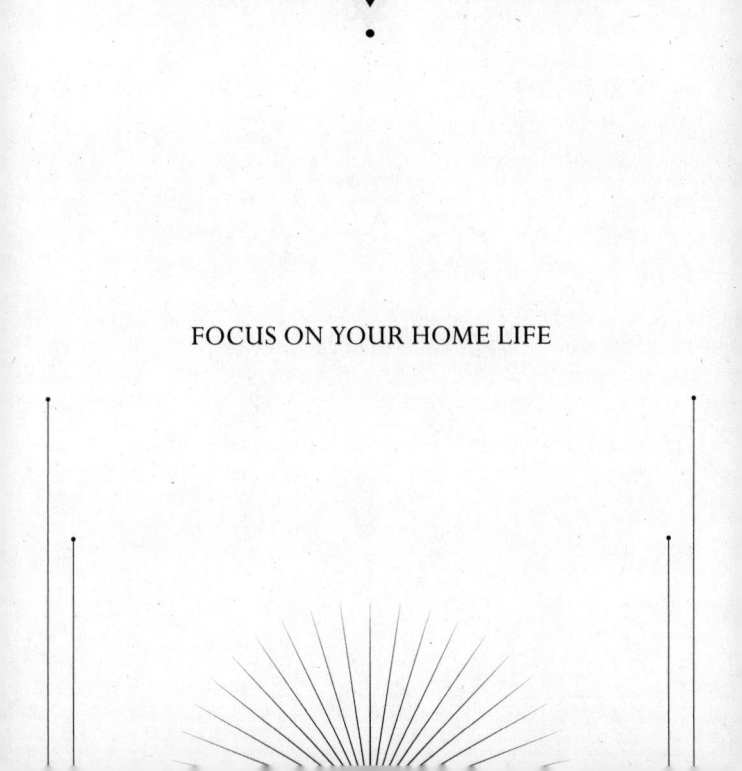

FOCUS ON YOUR HOME LIFE

专注于你的家庭生活

INVESTIGATE

AND THEN ENJOY IT

探索并享受它

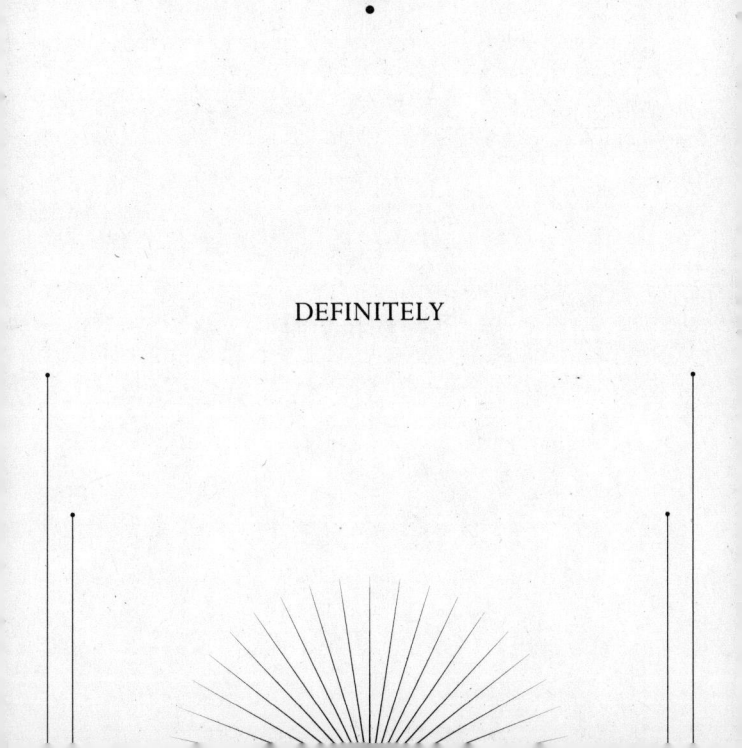

DEFINITELY

肯定地

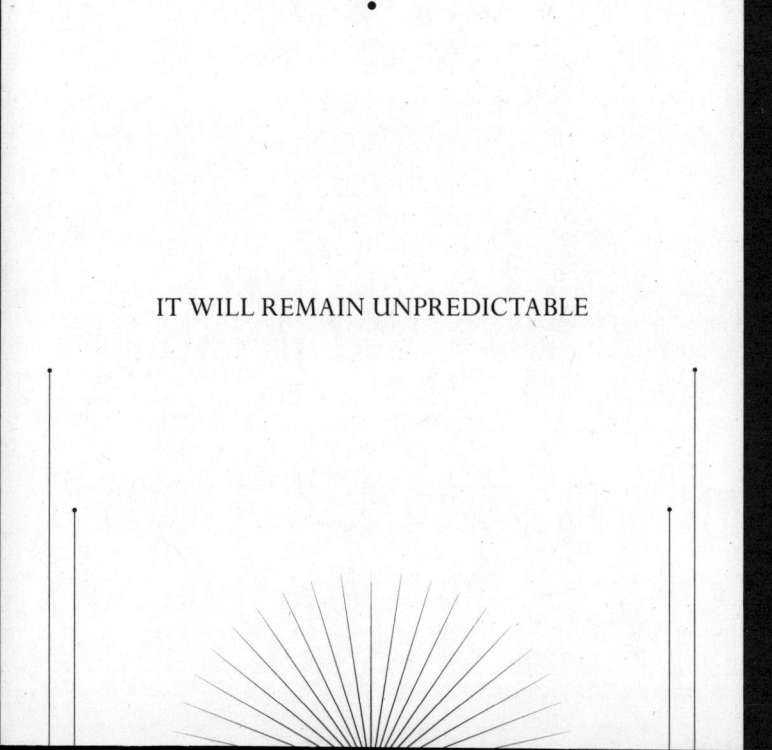

IT WILL REMAIN UNPREDICTABLE

仍然难以预料

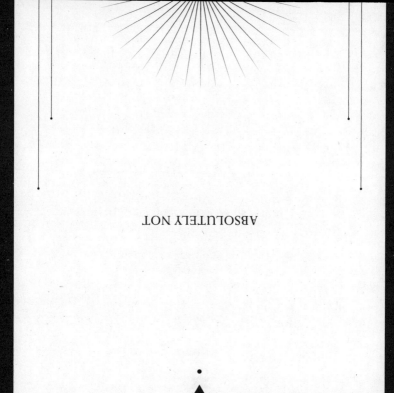

ABSOLUTELY NOT.

绝对不是

EXPLORE IT

WITH PLAYFUL CURIOSITY

好奇地去探索

BE DELIGHTFULLY SURE OF IT.

可以非常高兴地确定

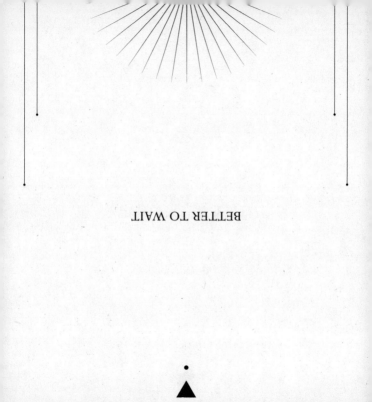

BETTER TO WAIT

最好等等

IT SEEMS ASSURED

似乎没有问题

DO IT EARLY

尽早行动

KEEP IT TO YOURSELF

不要告诉别人

STARTLING EVENTS

MAY OCCUR AS A RESULT

结果可能会令人吃惊

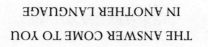

THE ANSWER COME TO YOU
IN ANOTHER LANGUAGE

答案可能会以另一种形式出现

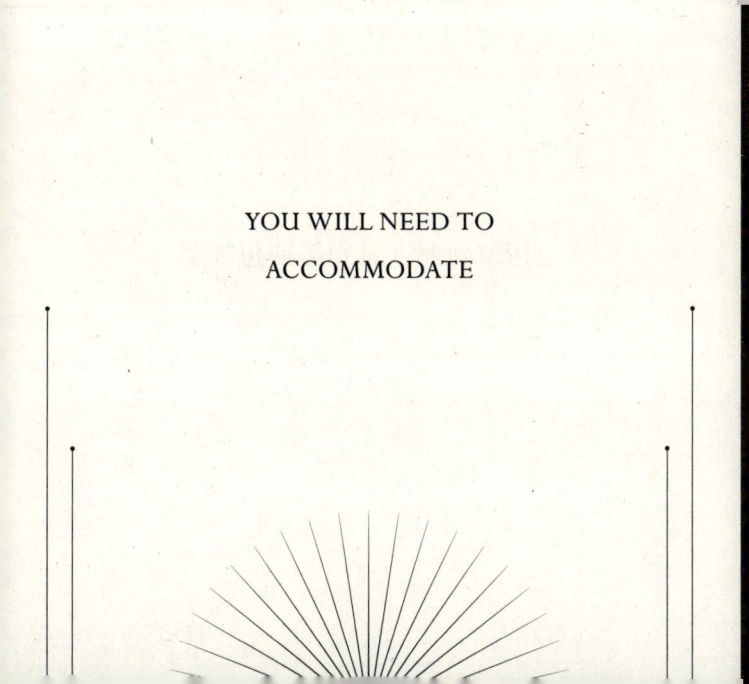

YOU WILL NEED TO

ACCOMMODATE

你需要适应

DOUBT IT

怀疑

IT WILL BRING GOOD LUCK

它会带来好运

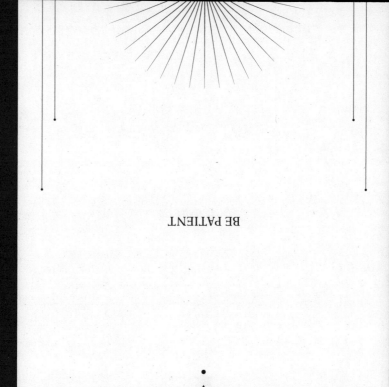

BE PATIENT

要有耐心

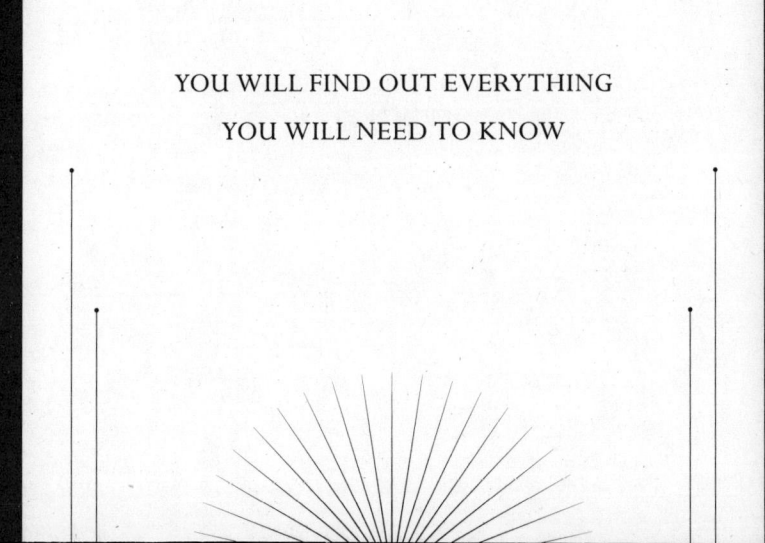

YOU WILL FIND OUT EVERYTHING

YOU WILL NEED TO KNOW

需要知道的，你都会知道

THERE IS A SUBSTANTIAL LINK

TO ANOTHER SITUATION

与另一种情况大有关系

WATCH AND SEE WHAT HAPPENS

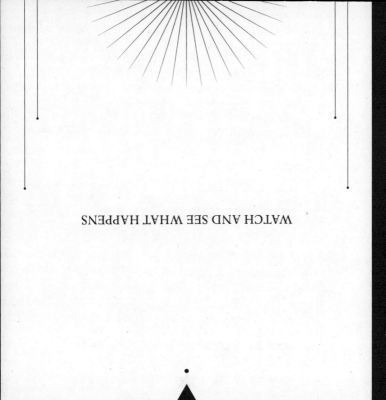

观察看看会发生什么

IT WILL AFFECT
HOW OTHERS SEE YOU

这会影响其他人对你的看法

YOU WILL BE HAPPY YOU DID

你会高兴你做了

GET IT IN WRITING

写下来

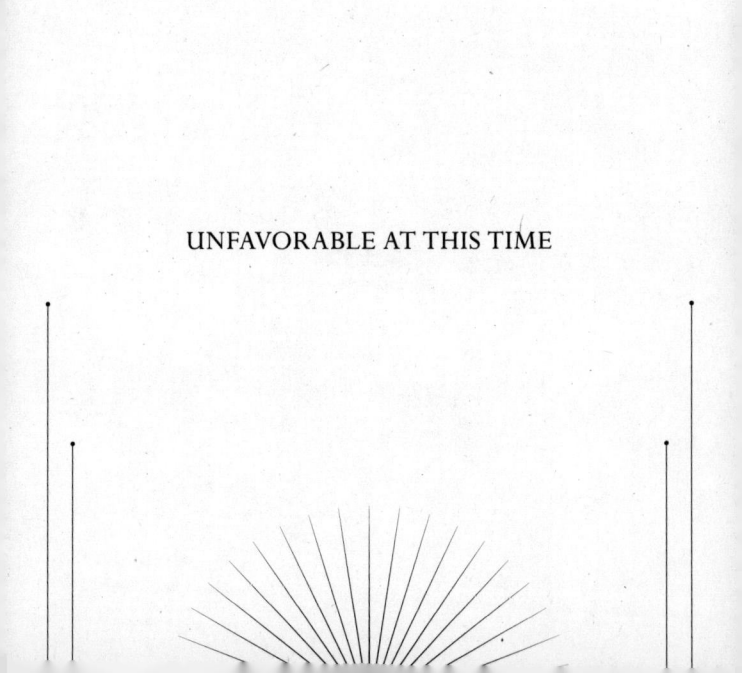

UNFAVORABLE AT THIS TIME

此时不宜

UPGRADE ANY WAY YOU CAN

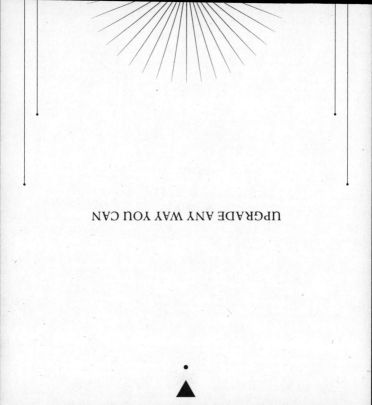

尽力提高

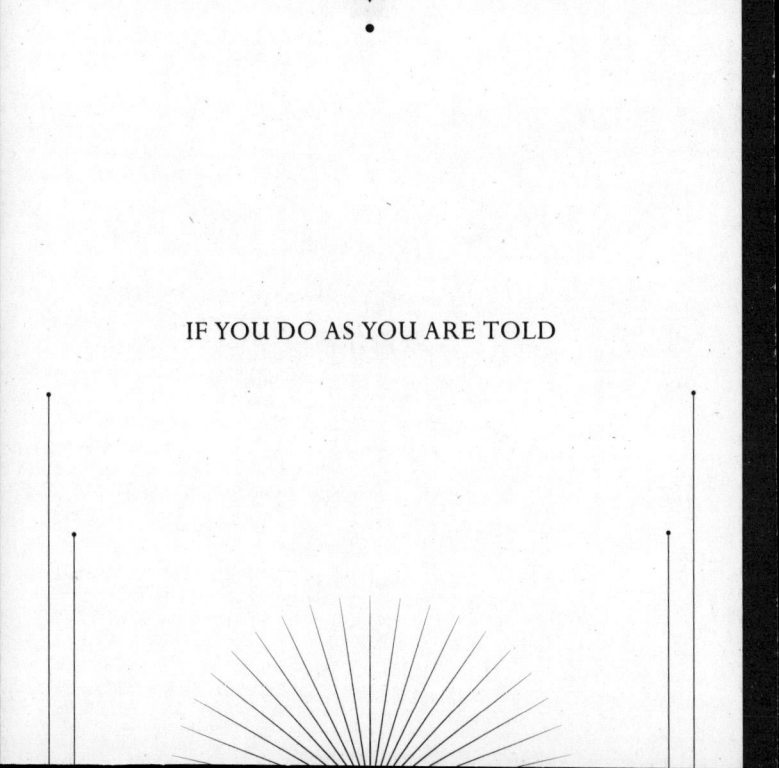

IF YOU DO AS YOU ARE TOLD

只要你按照被告知的方法去做

IF IT IS DONE WELL;

IF NOT, DO NOT DO IT AT ALL

能做好就做，否则干脆别做

DO NOT ASK FOR ANYMORE

AT THIS TIME

此时不要要求更多

AVOID THE FIRST SOLUTION

不要想到什么就做什么

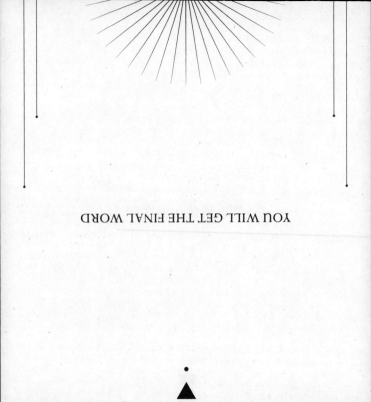

YOU WILL GET THE FINAL WORD

将由你来决定

PROCEED AT A MORE

RELAXED PACE

慢慢来

THE BEST SOLUTION

MAY NOT BE THE OBVIOUS ONE

轻易想到的办法未必是最好的

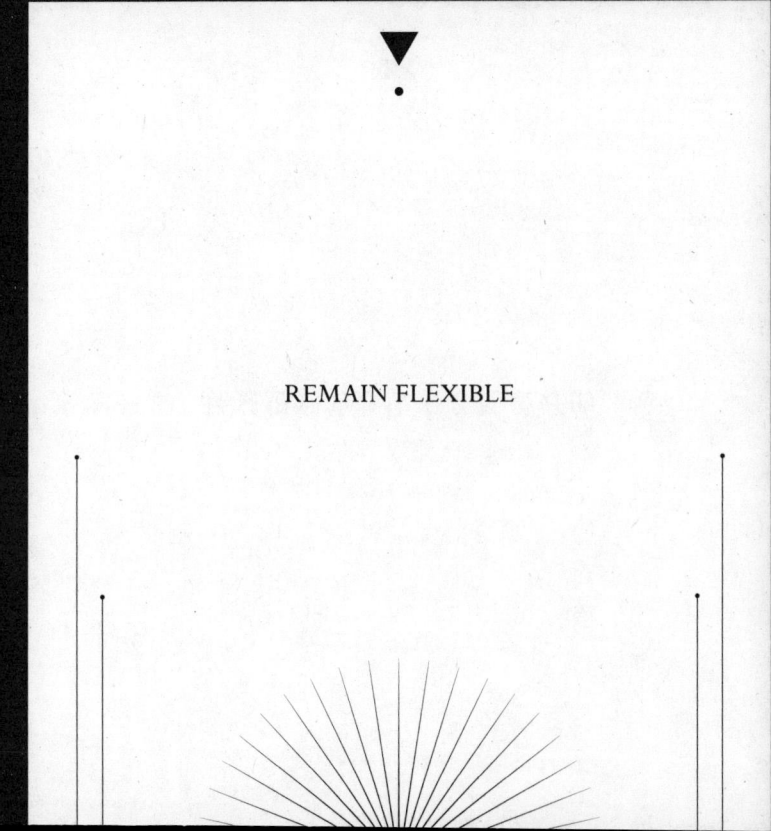

REMAIN FLEXIBLE

保持变通

THAT'S OUT OF YOUR CONTROL

超出你的掌控范围

PROVIDED YOU SAY

"THANK YOU."

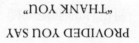

只要你说过"谢谢"

ENJOY THE EXPERIENCE

享受这种经历

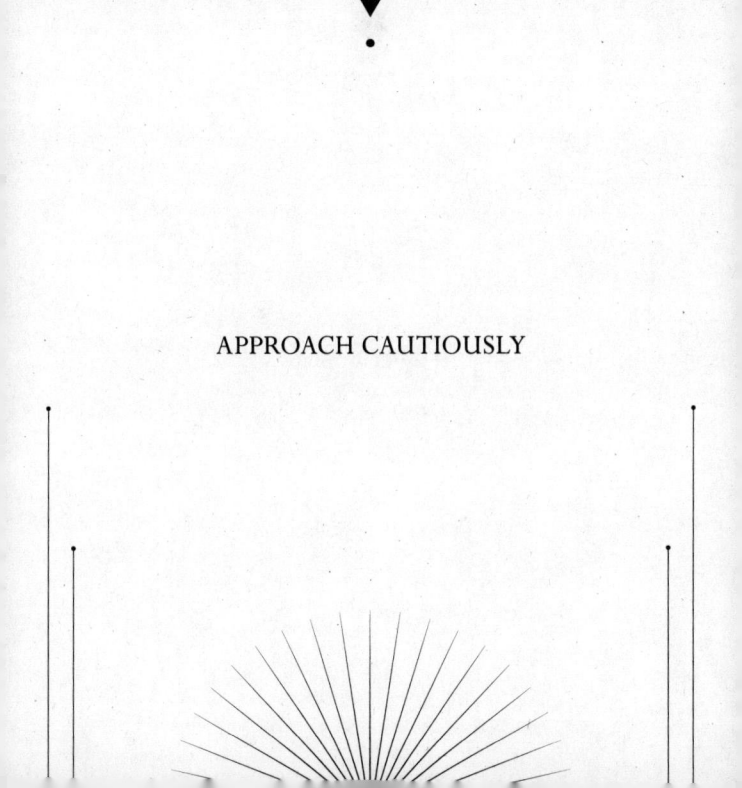

APPROACH CAUTIOUSLY

谨慎行事

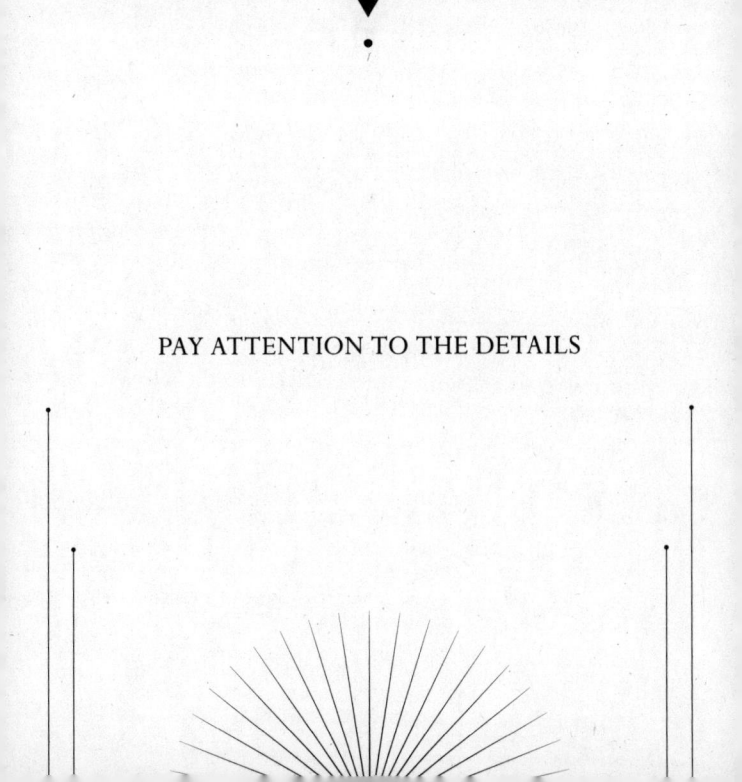

PAY ATTENTION TO THE DETAILS

注意细节

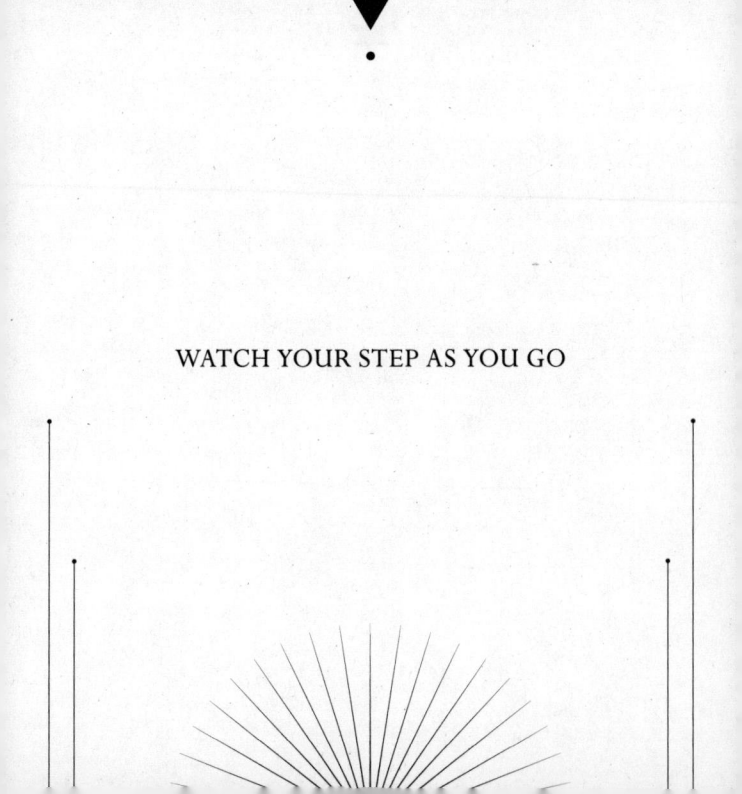

WATCH YOUR STEP AS YOU GO

留神走的每一步

SPEAK UP ABOUT IT

大声说出来

DO NOT HESTITATE

不要犹豫

THIS IS A GOOD TIME TO

MAKE A NEW PLAN

此时是制订新计划的好时机

MOVE ON

放下，然后继续

THERE IS NO GUARANTEE

无法保证

THE CIRCUMSTANCES

WILL CHANGE VERY QUICKLY

情况很快就会有所改变

DO NOT GET CAUGHT UP

IN YOUR EMOTIONS

不要被情绪左右

SHIFT YOUR FOCUS

转移你的焦点

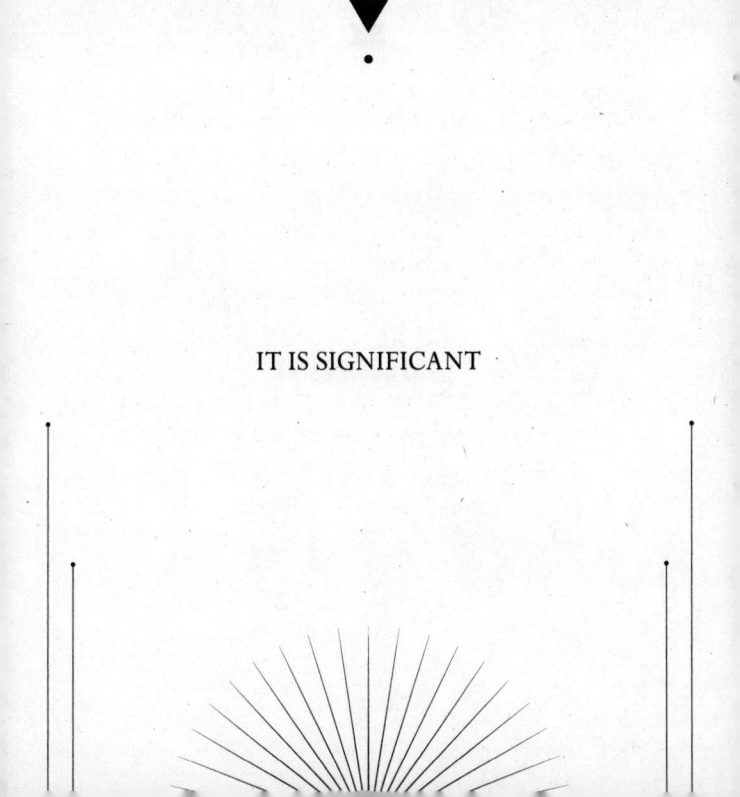

IT IS SIGNIFICANT

极为重要

REPRIORITIZE WHAT IS IMPORTANT

弄清事情的轻重缓急

MAKE A LIST OF WHY NOT

列出不做的理由

DO NOT WAIT

不要等待

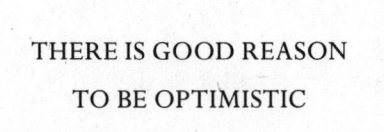

THERE IS GOOD REASON

TO BE OPTIMISTIC

有充分的理由保持乐观

IT IS SOMETHING

YOU WON'T FORGET

将使你难忘

ON

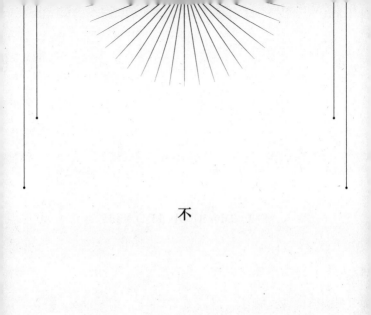

不

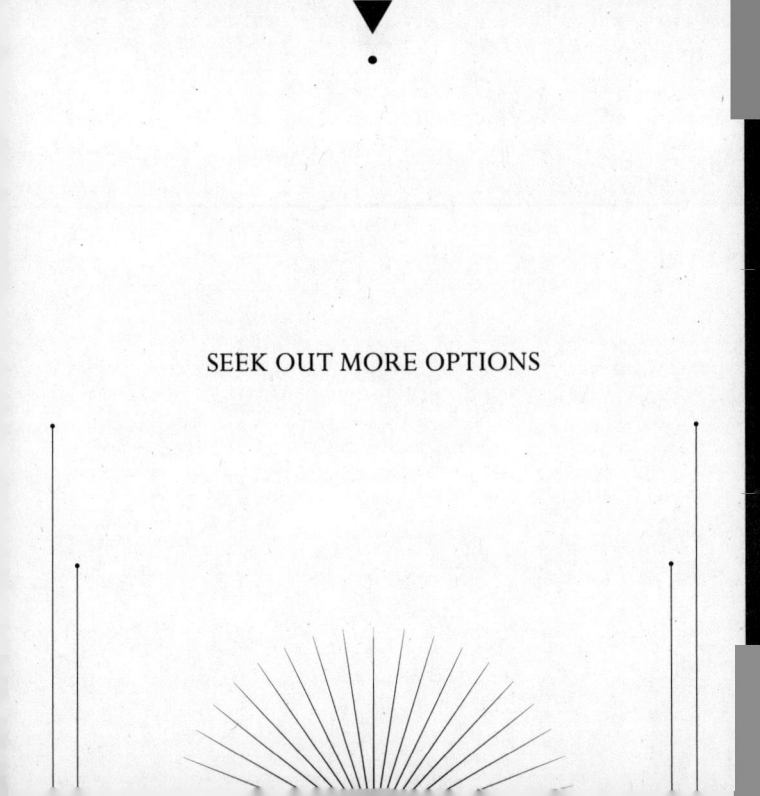

SEEK OUT MORE OPTIONS

寻求更多的选择

FOLLOW THROUGH
ON YOUR OBLIGATIONS

履行你的义务

DEAL WITH IT LATER

先不做决定

FOLLOW SOMEONE ELSE'S LEAD

跟着其他人走

MAKE A LIST OF WHY

列出做的理由

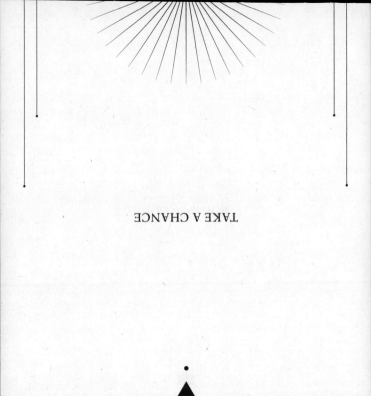

TAKE A CHANGE

冒险一试

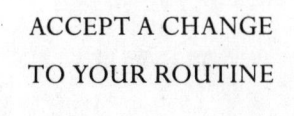

ACCEPT A CHANGE
TO YOUR ROUTINE

接受新的变化

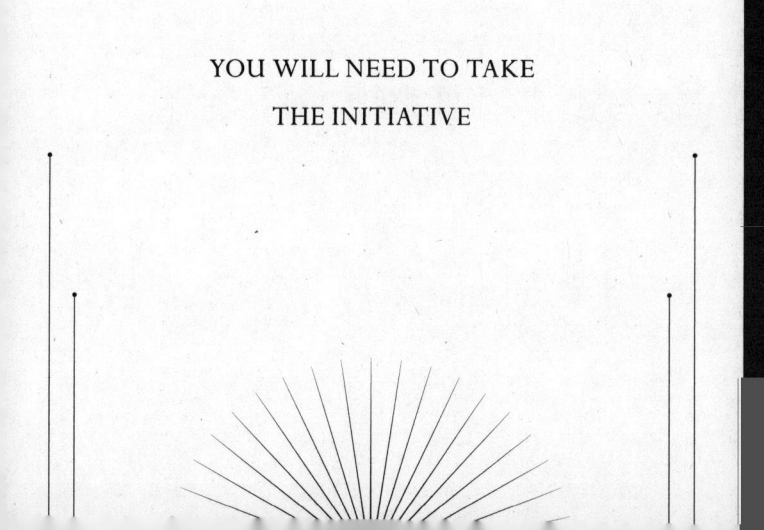

YOU WILL NEED TO TAKE

THE INITIATIVE

你需要争取主动

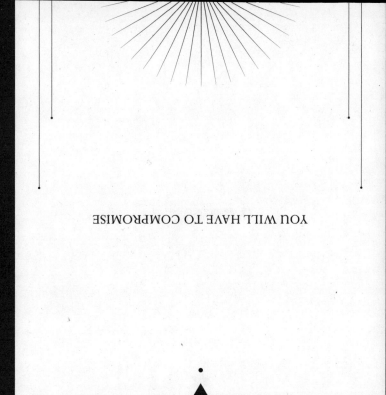

YOU WILL HAVE TO COMPROMISE

你必须妥协

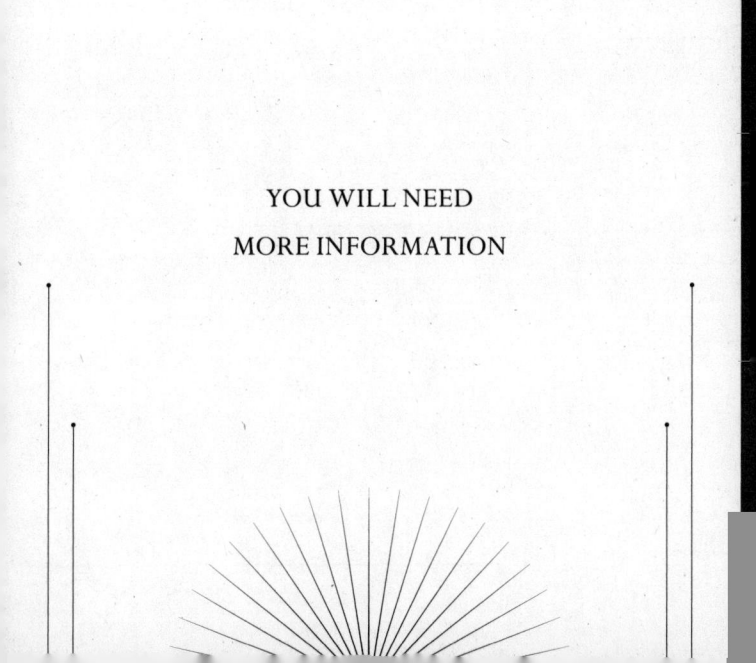

YOU WILL NEED

MORE INFORMATION

你需要更多的信息

TRUST
YOUR ORIGINAL THOUGHT

相信你最初的想法

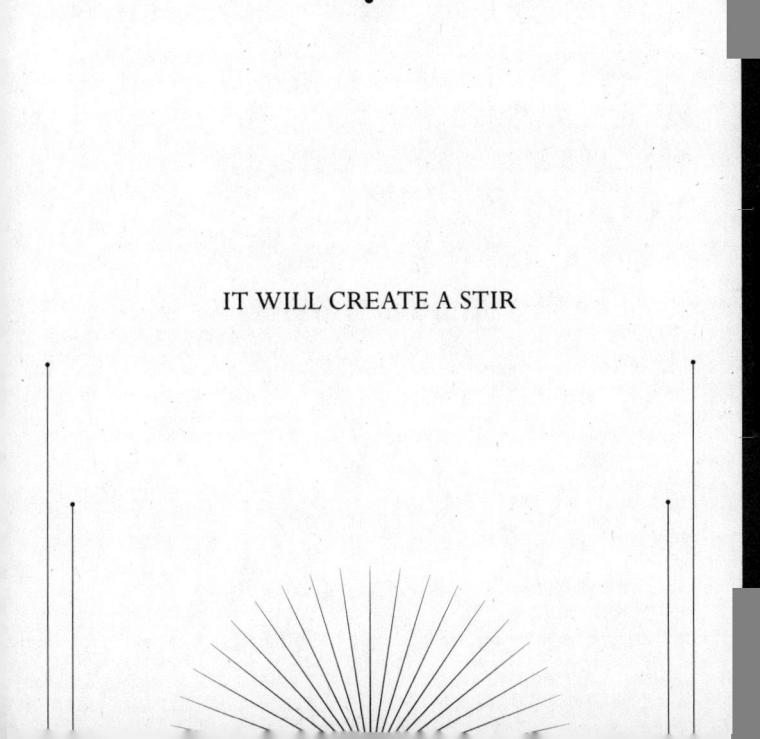

IT WILL CREATE A STIR

它会引起轰动

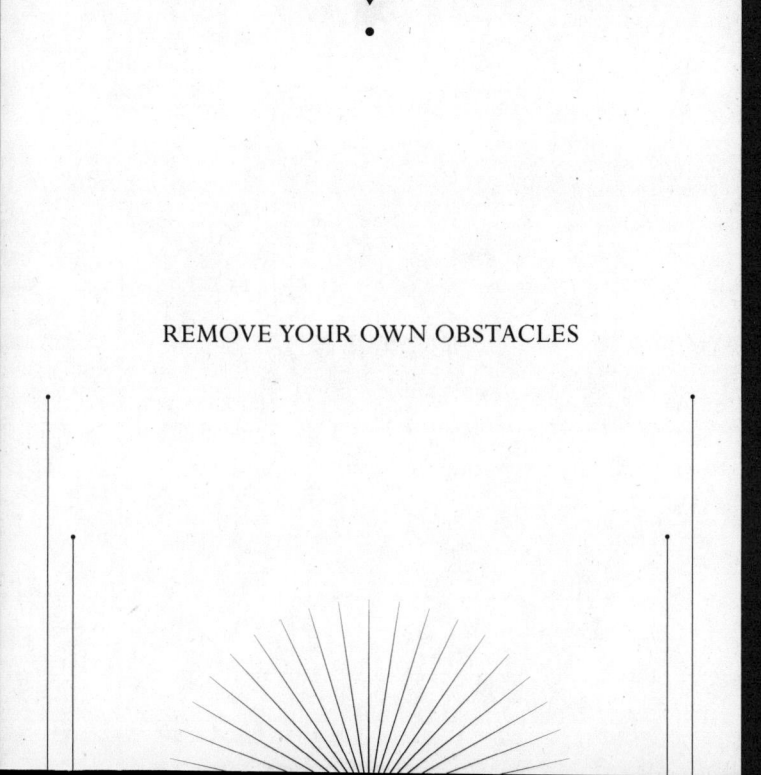

REMOVE YOUR OWN OBSTACLES

排除你自己的障碍

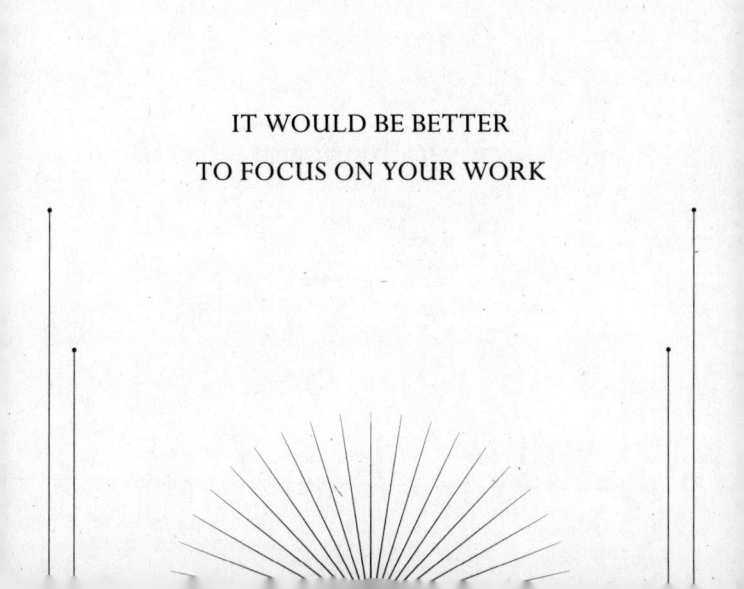

IT WOULD BE BETTER

TO FOCUS ON YOUR WORK

最好专注于你的工作

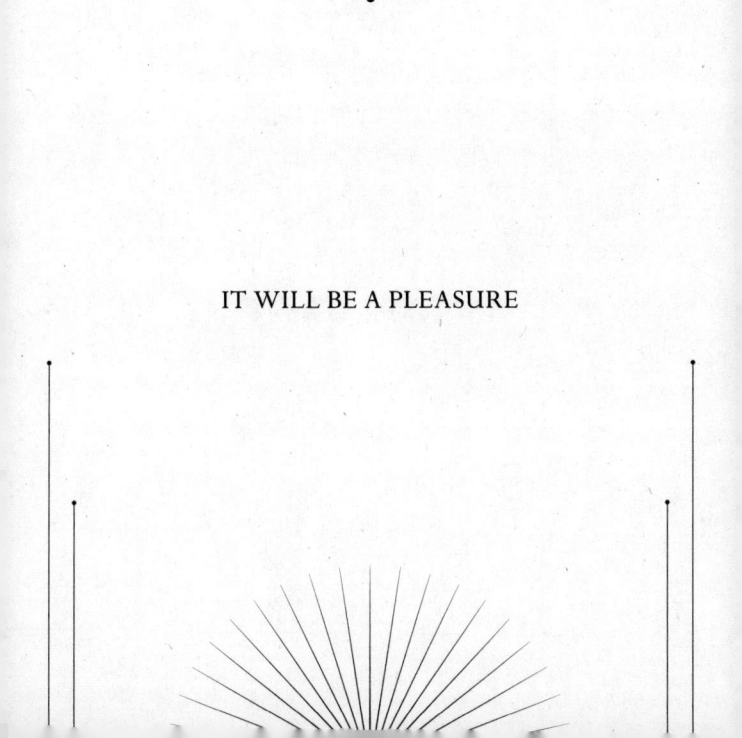

IT WILL BE A PLEASURE

将会很愉快

BE MORE GENEROUS

慷慨一点

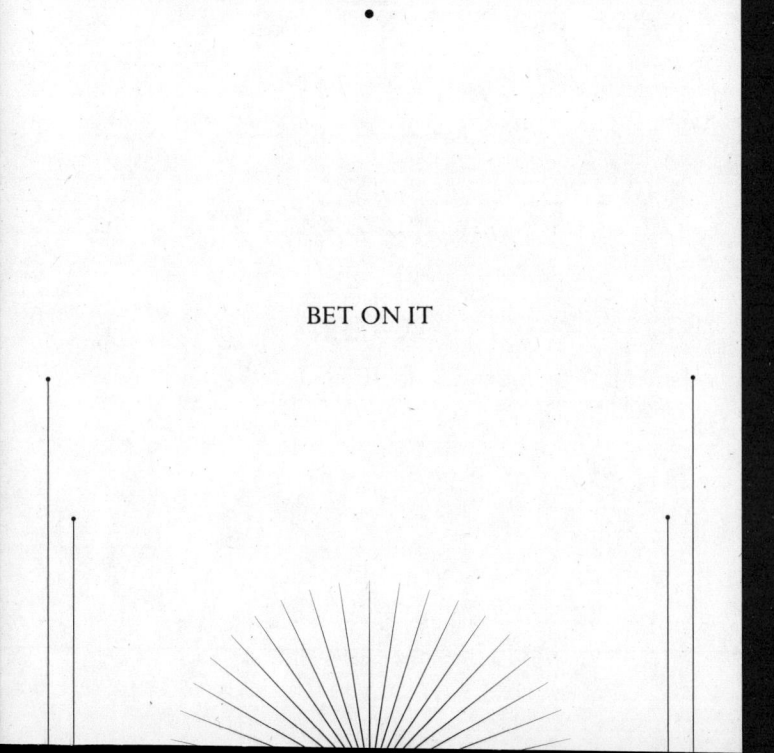

BET ON IT

十拿九稳

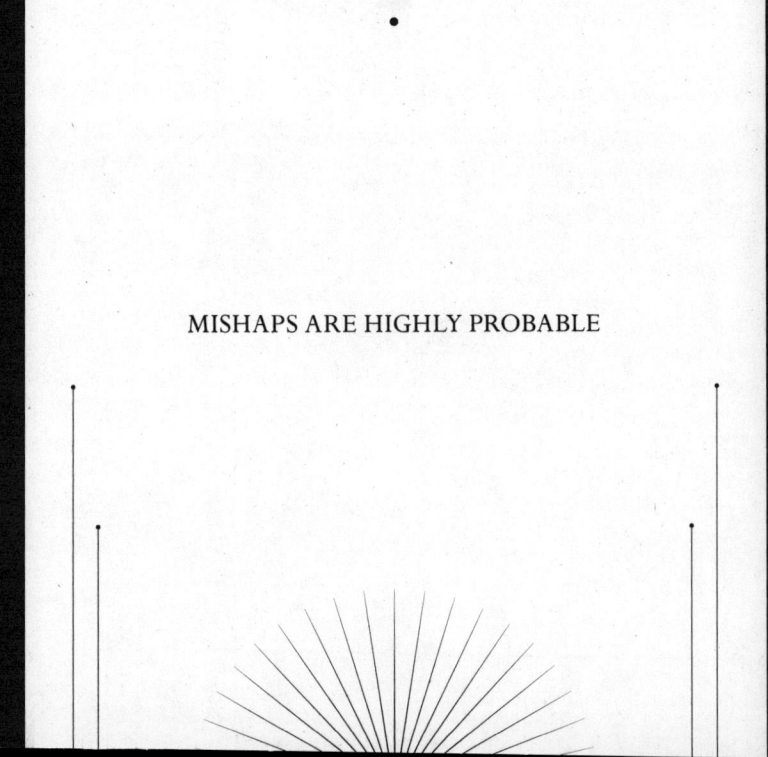

MISHAPS ARE HIGHLY PROBABLE

极有可能出现意外状况

PRESS FOR CLOSURE

尽快了结

REALIZE THAT

TOO MANY CHOICES IS AS

DIFFICULT AS TOO FEW

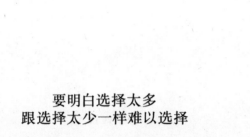

要明白选择太多
跟选择太少一样难以选择

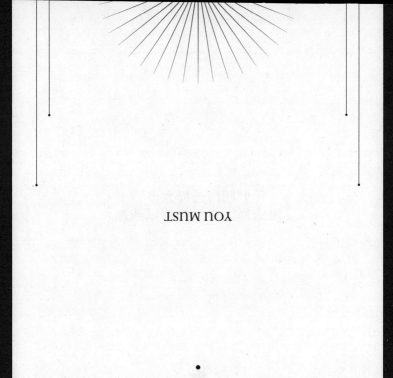

YOU MUST

你必须

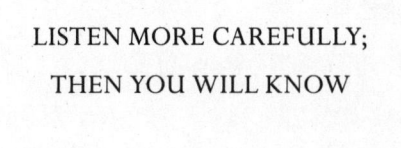

LISTEN MORE CAREFULLY;

THEN YOU WILL KNOW

仔细听，便会明白

THE ANSWER

IS IN YOUR BACKYARD

答案就在你内心深处

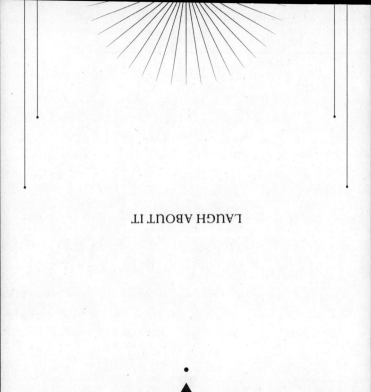

LAUGH ABOUT IT

一笑置之

OTHERS WILL DEPEND

ON YOUR CHOICES

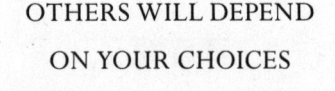

其余的，取决于你的选择

LET IT GO

随它吧

THAT WOULD BE A WASTE

OF MONEY

浪费钱

IT IS TIME FOR YOU TO GO

是该行动的时候了

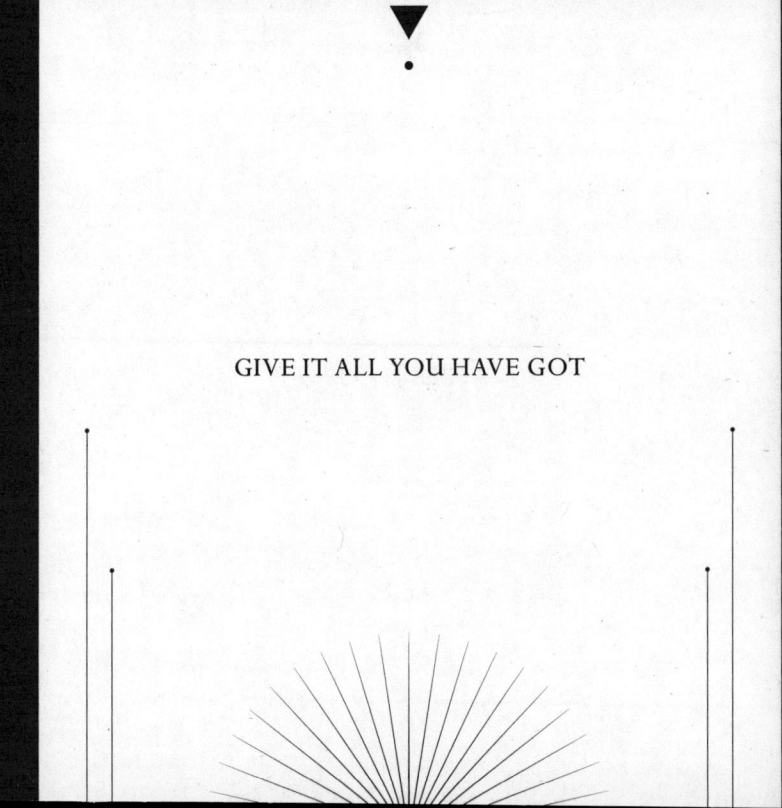

GIVE IT ALL YOU HAVE GOT

竭尽全力

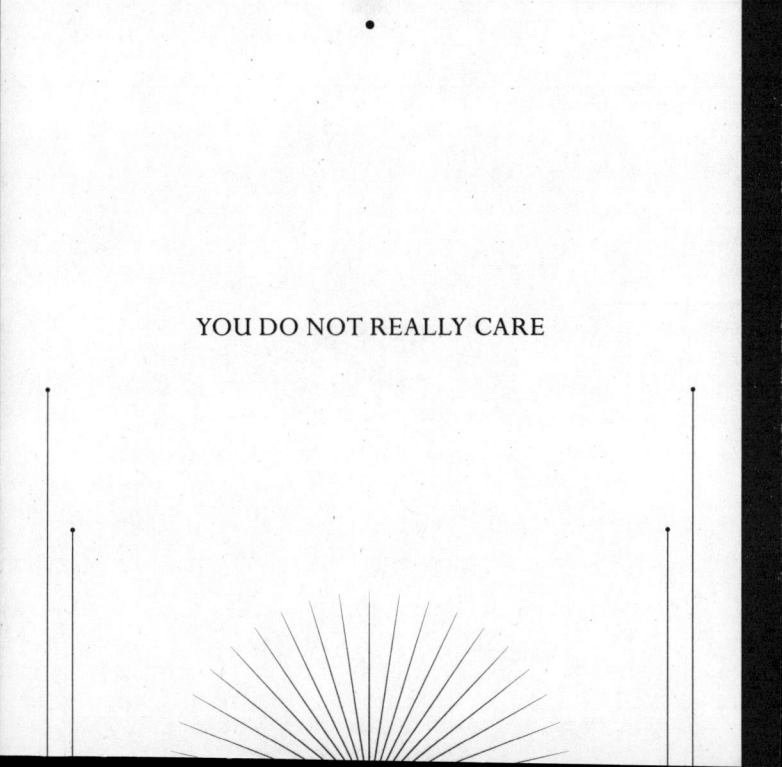

YOU DO NOT REALLY CARE

你并非真的在意

YOU WILL NEED TO
CONSIDER OTHER WAYS

你需要考虑其他办法

A YEAR FROM NOW

IT WILL NOT MATTER

一年之后这将不再重要

DO NOT WASTE YOUR TIME

不要浪费你的时间

IT COULD BE EXTRAORDINARY

可能会异乎寻常地好

COUNT TO 10;
ASK AGAIN

从 1 数到 10，再问一次

ACT AS THOUGH
IT IS ALREADY REAL

就当它是既成事实

SETTING PRIORITIES

WILL BE A NECESSARY PART

OF THE PROCESS

弄清轻重缓急是此事的重中之重

USE YOUR IMAGINATION

发挥你的想象力

IT IS GONNA BE GREAT

一定会非常成功

TO ENSURE THE BEST DECISION,

BE CALM

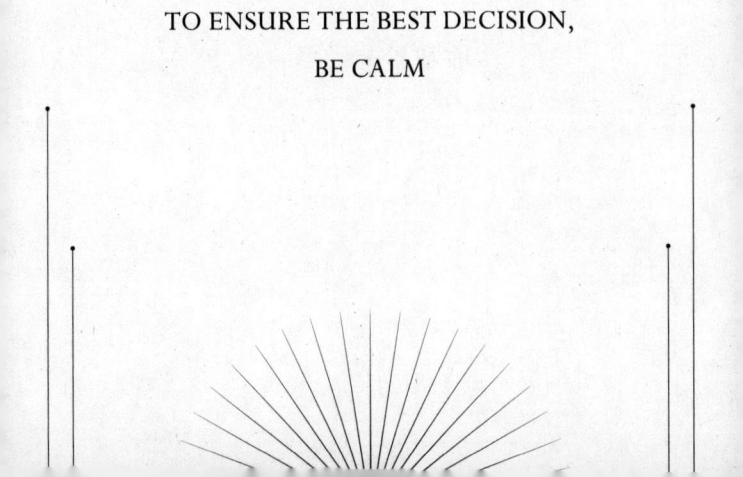

冷静才能做出最好的选择

WAIT

等待

YOU WILL HAVE TO MAKE IT UP

AS YOU GO

你到时将不得不随机应变

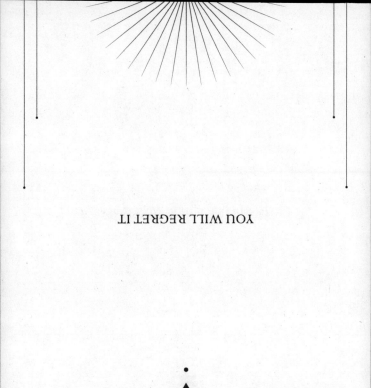

YOU WILL REGRET IT

你会后悔的

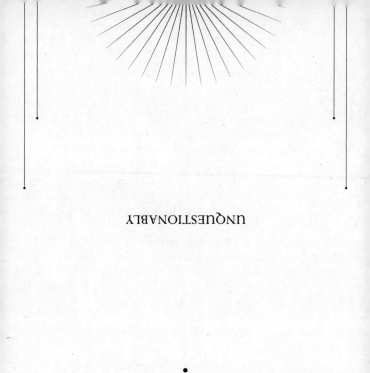

UNQUESTIONABLY

毫无疑问

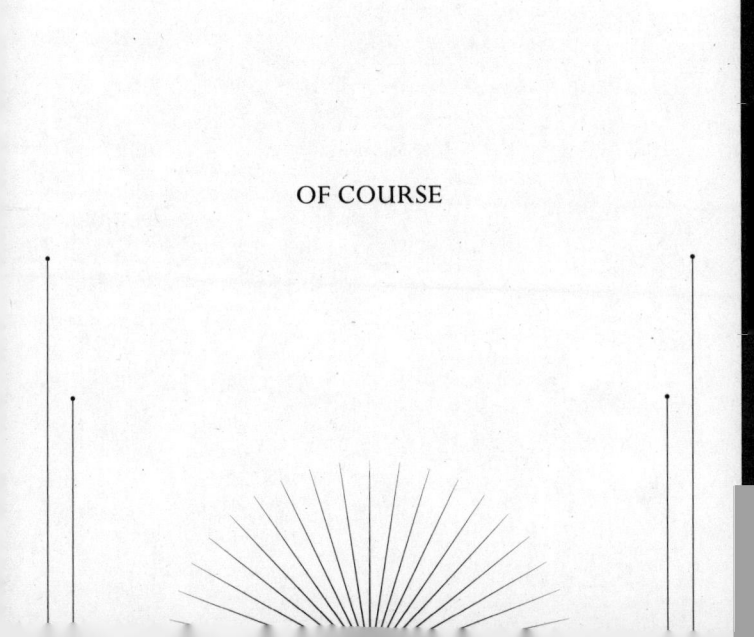

OF COURSE

当然

YOU KNOW BETTER NOW
THAN EVER BEFORE

你现在比以前任何时候都更清楚

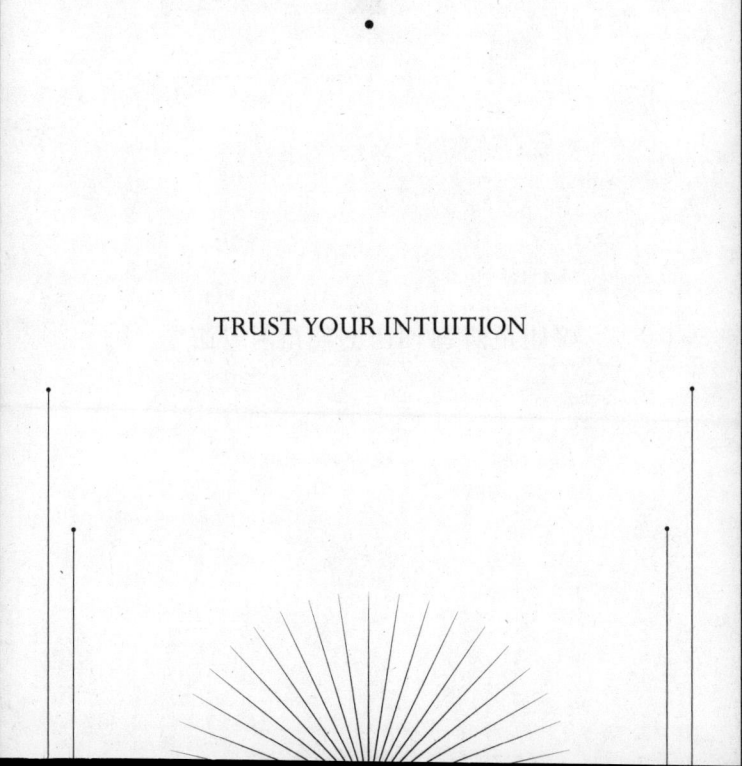

TRUST YOUR INTUITION

相信你的直觉

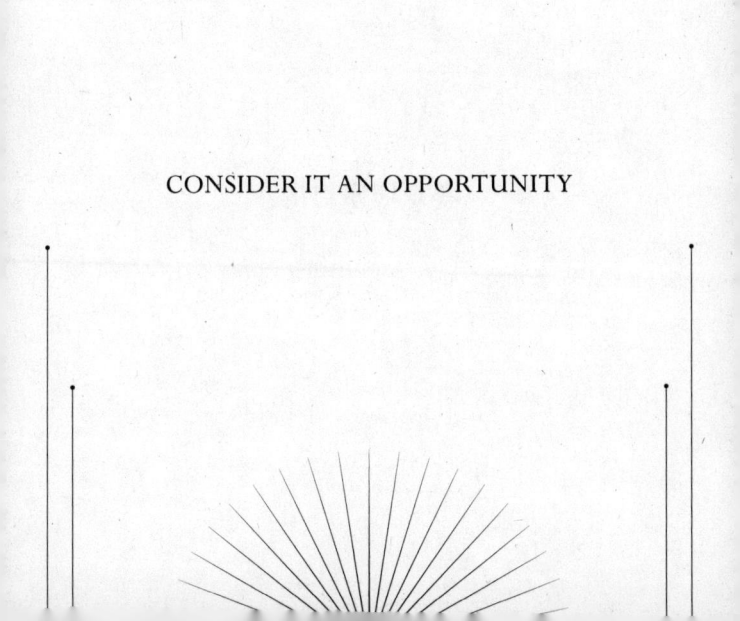

CONSIDER IT AN OPPORTUNITY

把它看作一个机会

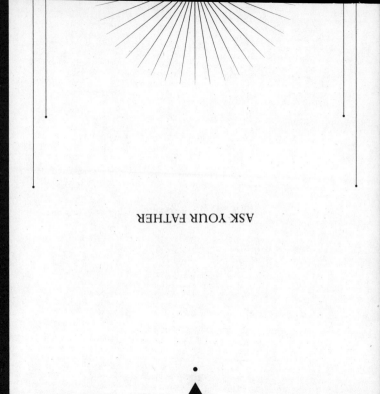

ASK YOUR FATHER

问问父亲

NEVER

绝不

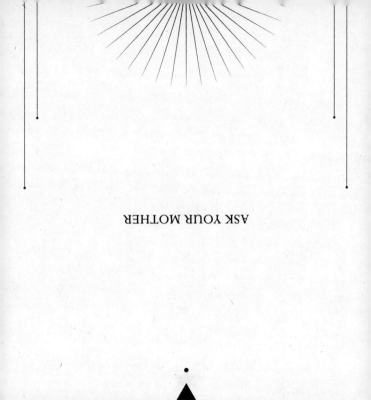

ASK YOUR MOTHER

问问母亲

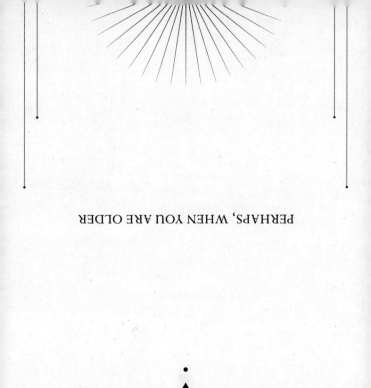

PERHAPS, WHEN YOU ARE OLDER

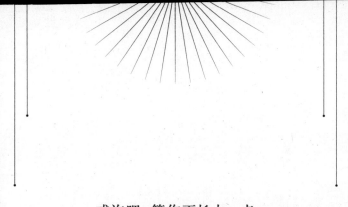

或许吧，等你再长大一点

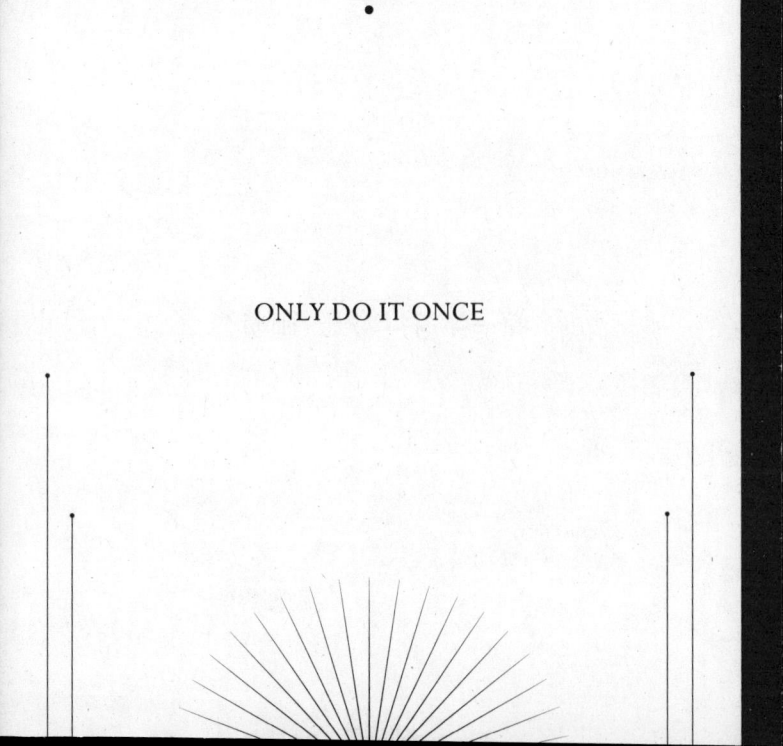

ONLY DO IT ONCE

仅此一次

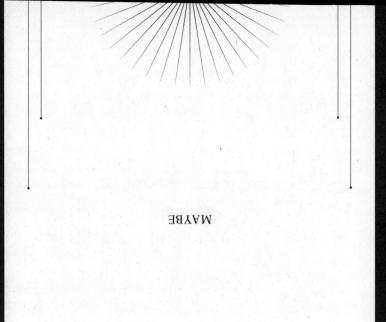

MAYBE

也许吧

ON

不

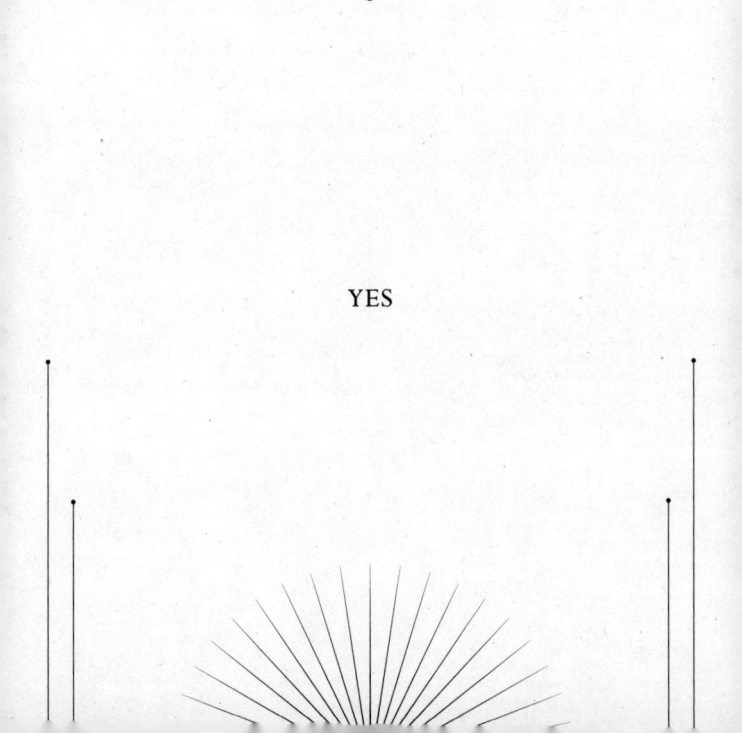

YES

是的

FINISH SOMETHING ELSE FIRST

先完成其他的事情

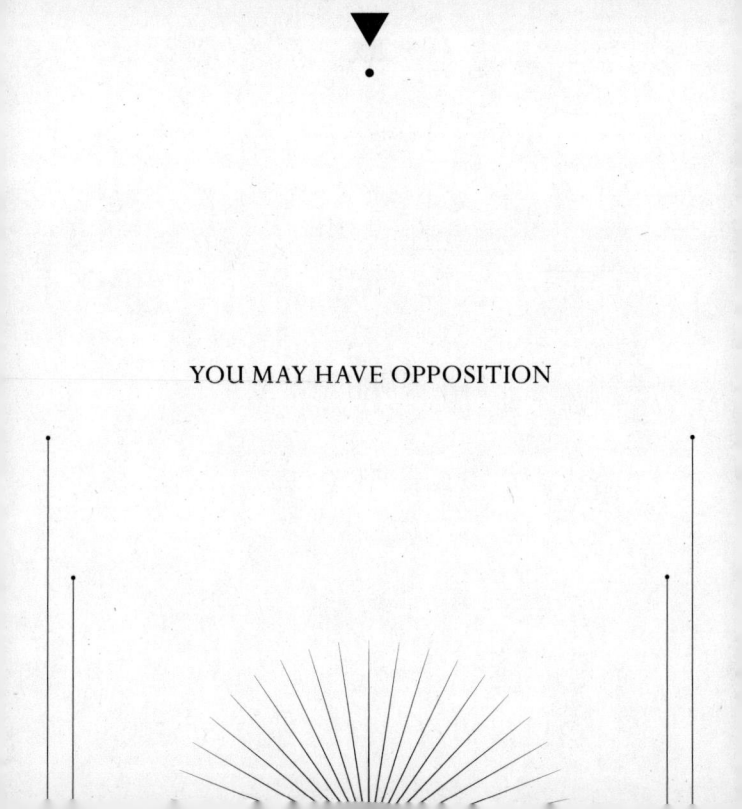

YOU MAY HAVE OPPOSITION

你可能遭到反对

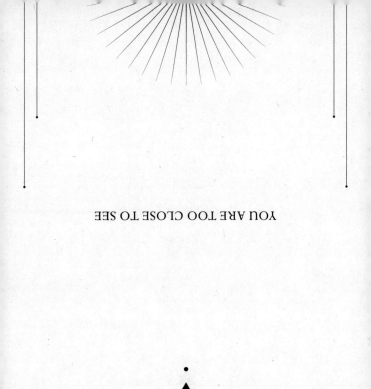

YOU ARE TOO CLOSE TO SEE

你离得太近，难以看清

THE SITUATION IS UNCLEAR

情况不明

A SUBSTANTIAL EFFORT

WILL BE REQUIRED

需要付出巨大努力

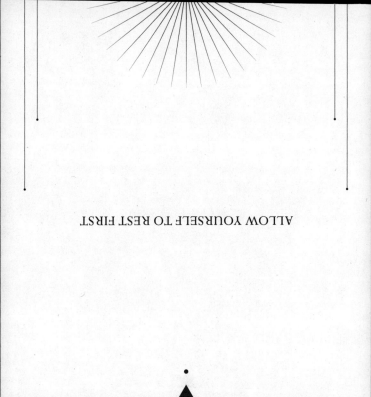

ALLOW YOURSELF TO REST FIRST

让自己先休息一下

THE CHANCE

WILL NOT COME AGAIN SOON

这个机会很难得

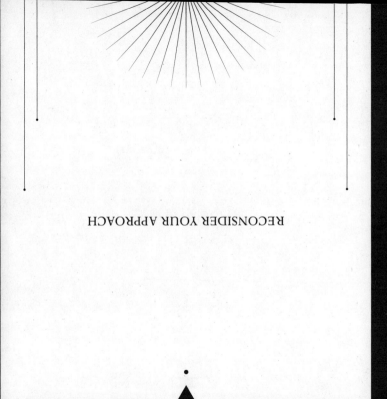

RECONSIDER YOUR APPROACH

重新考虑你的方法

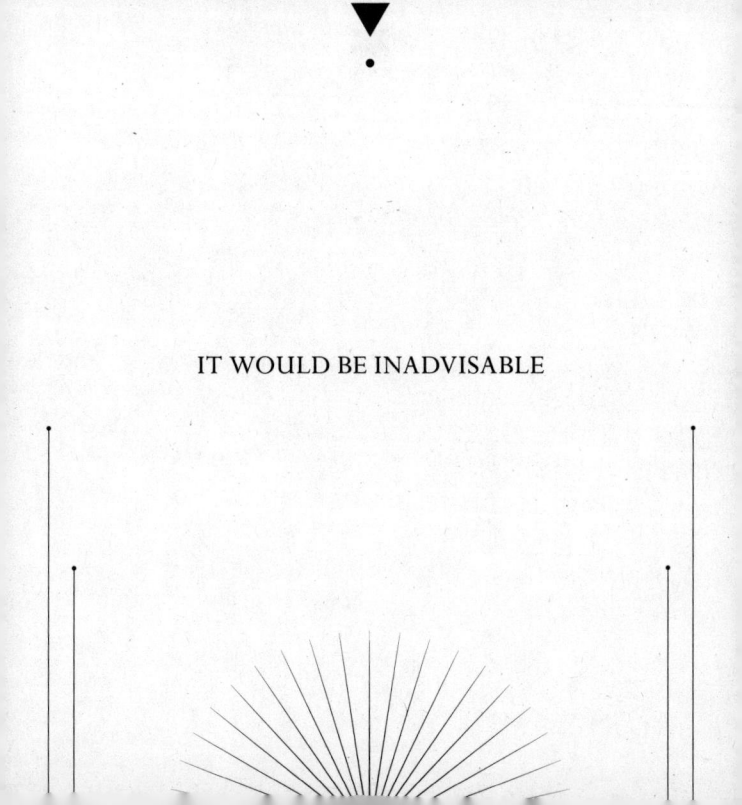

IT WOULD BE INADVISABLE

不妥

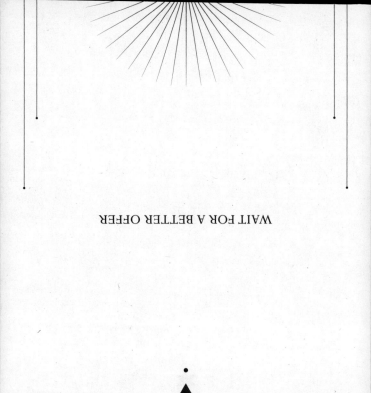

WAIT FOR A BETTER OFFER

等待更好的机会

SETTLE IT SOON

尽快解决

YES,

BUT DON'T FORCE IT

是的，但不要强迫

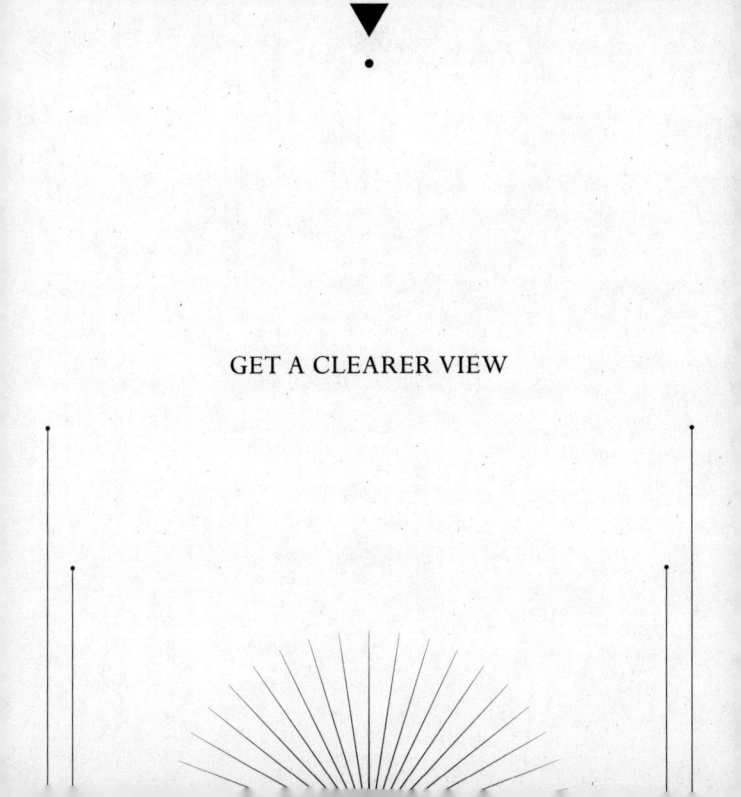

GET A CLEARER VIEW

看得更清楚一点

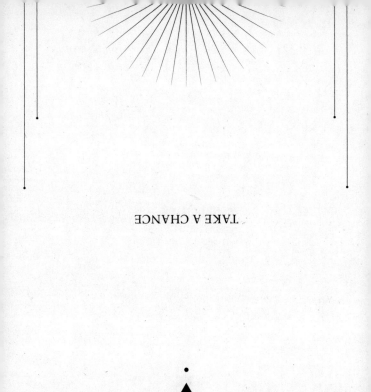

TAKE A CHANCE

冒险一试

NOW YOU CAN

现在可以了

DON'T OVERDO IT

不可过分

IT WILL SUSTAIN YOU

它会让你受益

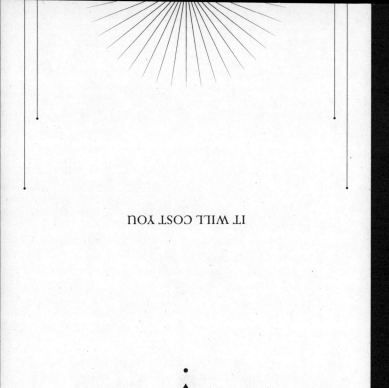

IT WILL COST YOU

它会使你付出

IT IS SURE TO MAKE THINGS
INTERESTING

这会使事情变得更有趣

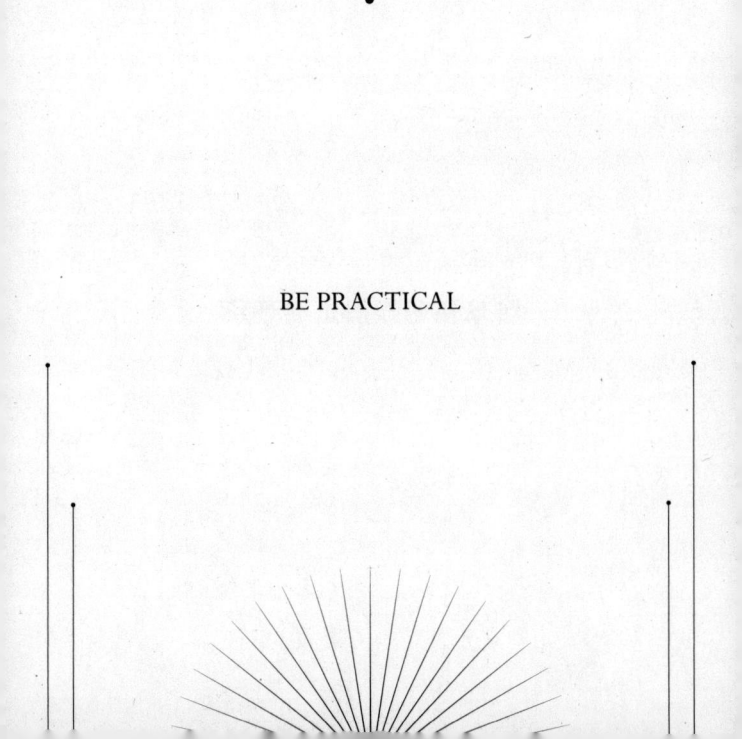

BE PRACTICAL

实际点

SAVE YOUR ENERGY

保存你的精力

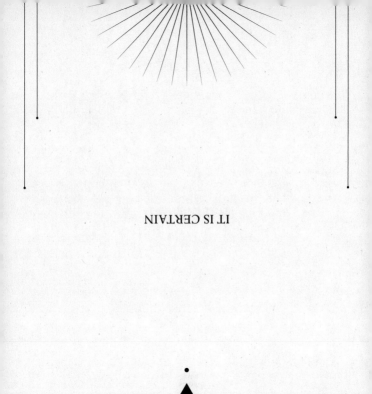

IT IS CERTAIN

确定无疑

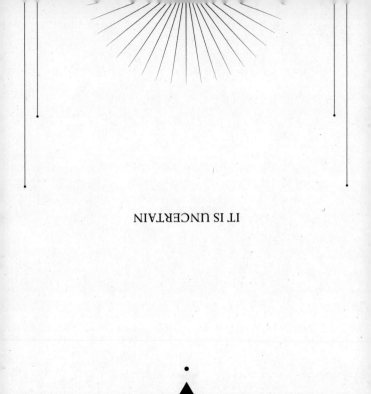

IT IS UNCERTAIN

不确定

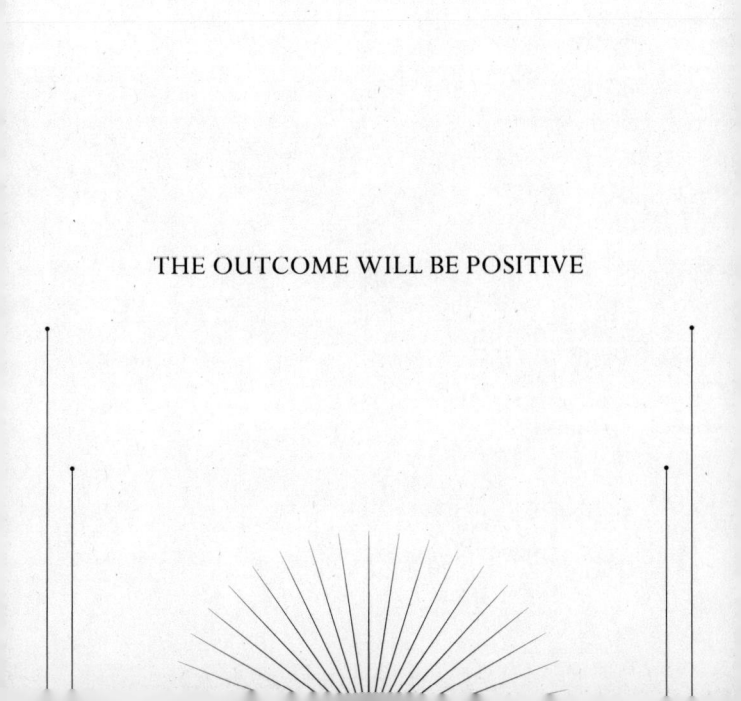

THE OUTCOME WILL BE POSITIVE

结果将是好的

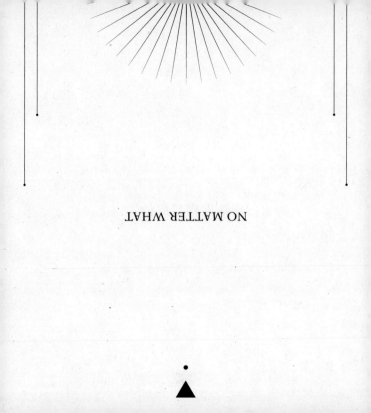

NO MATTER WHAT

哪怕天上下刀子

致 谢

《答案之书》自创作之初就得到了很多朋友的支持,因而我难以在这短短的一页中将我想感谢的人一一列举出来。然而,有些人对我的支持和对本书的创作确实做出了特殊的贡献,我必须在此表示感谢。

我的母亲多丽丝和父亲鲍伯总能为我提供正确的答案,谢谢你们。

我的知己克里斯对稿件进行了后期加工,我在此对他深表谢意。

同时感谢我的朋友桑德拉的贡献和持续的支持。

佩格和拉里与我共同度过了一段真诚而美好的时光，感谢他们的慷慨支持。

我的著作经理人维多利亚·桑德斯热情洋溢，有极好的判断力，感谢她的信任和自始至终的支持。

感谢我的编辑詹妮弗·兰，她帮我处理所有烦琐的细节而热情不减。

还有我在这里无法一一致谢的西雅图的众多支持者，特别是扎格、蒂姆、玛塔、迈克尔、约书亚、芭芭拉、莫林和瑞妮，以及在轮船咖啡馆的朋友们。

文景

社科新知 文艺新潮

Horizon

答案之书(双语精巧版)

〔加拿大〕卡罗尔·博尔特 著
何静 译

出 品 人:姚映然
责任编辑:王 萌
营销编辑:高晓倩
装帧设计:王柿原

出 品:北京世纪文景文化传播有限责任公司
(北京朝阳区东土城路8号林达大厦A座4A 100013)
出版发行:上海人民出版社
印 刷:天津联城印刷有限公司

开本:960mm×1092mm 1/64
印张:10 字数:30,000
2022年1月第1版 2025年1月第16次印刷
定价:38.00元
ISBN:978-7-208-17442-9 / B·1591

图书在版编目(CIP)数据

答案之书:双语精巧版:汉、英/(加)卡罗尔·
博尔特(Carol Bolt)著;何静译. —— 上海:上海人
民出版社,2021
书名原文:The Book of Answers
ISBN 978-7-208-17442-9

Ⅰ.①答… Ⅱ.①卡…②何… Ⅲ.①人生哲学-通
俗读物-汉、英 Ⅳ.①B821-49

中国版本图书馆CIP数据核字(2021)第224179号

本书如有印装错误,请致电本社更换 010-52187586